Computational Biology and Chemistry

Edited by Payam Behzadi and Nicola Bernabò

Published in London, United Kingdom

IntechOpen

Supporting open minds since 2005

Computational Biology and Chemistry
http://dx.doi.org/10.5772/intechopen.83539
Edited by Payam Behzadi and Nicola Bernabò

Contributors
Muraleedharan Karuvanthodi, Mohamed Shahin Thayyil, Safna Hussan Kodakkat Parambil, Shameera Ahamed T.K., Jorge Álvarez Cervantes, Edna María Hernández Domínguez, Gerardo Diaz-Godinez, Yarely Garcia Esquivel, Virginia Mandujano-Gonzalez, Laura Sofía Castillo Ortega, Joji M. Otaki, Shiho Endo, Masakazu Tsuhako, Kenta Motomura, Yuuki Kakazu, Morikazu Nakamura, Songul Yasar Yildiz, Arthur Eugen Kümmerle, Thiago Moreira Pereira, Chin Lung Lu, Yi-Kung Shieh, Shu-Cheng Liu, Payam Behzadi

First published in London, United Kingdom, 2021 by IntechOpen
IntechOpen is the global imprint of INTECHOPEN LIMITED, registered in England and Wales, registration number: 11086078, 5 Princes Gate Court, London, SW7 2QJ, United Kingdom
Printed in Croatia

British Library Cataloguing-in-Publication Data
A catalogue record for this book is available from the British Library

Additional hard and PDF copies can be obtained from orders@intechopen.com

Computational Biology and Chemistry
Edited by Payam Behzadi and Nicola Bernabò
p. cm.
Print ISBN 978-1-78985-366-7
Online ISBN 978-1-78985-691-0
eBook (PDF) ISBN 978-1-78985-692-7

For EU product safety concerns:
IN TECH d.o.o., Prolaz Marije Krucifikse Kozulić 3, 51000 Rijeka, Croatia, info@intechopen.com or visit our website at intechopen.com.

Meet the editors

Dr. Payam Behzadi was born in 1973, in Tehran, Iran. He began his collaboration with the Department of Microbiology, College of Basic Sciences, Shahr-e-Qods Branch, Islamic Azad University as a faculty member (with an MSc degree in Microbiology) in 2004. He finally got his Ph.D. degree in Molecular biology in 2016 (BSc and MSc in Microbiology; Ph.D. in Molecular Biology) and now continues his scientific activities in the position of Assistant Professor at the same university. He has several students from different academic levels including BSc, MSc, and Ph.D. Dr. Payam Behzadi has authored and edited more than 20 chapters and academic books and more than 70 original and review articles. His scientific research interests are: urinary tract infections, antibiotics, bioinformatics, genetics, gene profiling, molecular biology and cellular and molecular immunology. Dr. Payam Behzadi trains as an ice skater in his free time.

Prof. Nicola Bernabò is an Associate Professor of Veterinary Physiology at the Faculty of Veterinary Medicine, University of Teramo, Italy. He received his degree in veterinary medicine from the University of Pisa (1999), his Master's degree in biotechnology of reproduction from the University of Teramo (2002), and Ph.D. degree in endocrinology of domestic animals from the University of Bologna (2004). His research interests include endocrinology of reproduction, the biochemistry of sperm capacitation, and systems biology. In particular, his work is focused on the development of computational models representing the complex biochemical events that occur during the acquisition of fertilizing ability by male gametes either in domestic animals or humans. Prof. Bernabò is the author of more than sixty works published in international peer-reviewed journals, conference proceedings, as well as book chapters, and he has been awarded national and international prizes for his research activity.

Contents

Preface

The employment of computers, software tools, and internet services in basic sciences and medicine has led to changes in investigation methodologies. This occurrence resulted in the establishment of multidisciplinary sciences involving bioinformatics and systems biology. The computational multidisciplinary sciences represented new types of studies and laboratories. In silico investigations and dry laboratories have been the invaluable products of these sciences. The serious and continuous activities in this regard led to the accumulation of a huge amount of digital data (bioinformatic data) in the form of databases. These progressions and facilities resulted in invaluable outcomes such as computational biology and chemistry. I call them soft biology and chemistry.

The reason for the success and progression in soft biology and chemistry is the appearance of effective and precise bioinformatic software tools and databases such as the National Center for Biotechnology Information (NCBI) (https://www.ncbi.nlm.nih.gov/), Kyoto Encyclopedia of Genes and Genomes (KEGG) (https://www.genome.jp/kegg/), and Research Collaboratory for Structural Bioinformatics PDB (RCSBPDB) (https://www.rcsb.org/).

Visualization and 3-dimentionization of different biomolecules, structures, and complexes are the most fantastic facilities that are provided and represented by bioinformatics, systems biology, computational biology, and chemistry.

The results of traditional investigations within wet labs by molecular biologists, biologists, biochemists, and chemists produced only raw data including nucleotides and amino acid sequences.

With the appearance of bioinformatics and the use of computational technologies, we are able to visualize these data to have their putative spatial configurations and conformations. Today, we are able to 3-dimensionalize the discovered raw data to have a limited imaging capability of their natural structure to understand the related characteristics, practicalities, and functionalities.

Moreover, the use of computational biology and chemistry has had effective consequences in pharmaceutics. As we know, the traditional procedure of preparation and provision of a drug or antibiotic includes several control and check processes in different levels of assays. These procedures and processes cost a lot of money and takes a long duration of time.

Recruitment of computational biology and chemistry represents a great opportunity for pharmaceutics. Drug designing is a brilliant bonus to pharmaceutics to shorten technical procedures and duration of time to reduce the costs of drugs and antibiotics.

All in all, we are at the beginning point of the soft biology and chemistry pathway. The progressions of these scientific disciplines support us to have incredible imaging, illustration, and interpretation of the obtained raw data in traditional wet labs. In the future we can have a precise image of the reason of spatial architecture of different biomolecules and their interactions with other structures and complexes. The dry labs determine the future of biology and chemistry!

The book "Computational Biology and Chemistry" is a collection of invaluable results and outcomes obtained by global and international scientists. This book involves seven chapters in five sections.

The first chapter is written by the editor and it completes the first section of the book. This chapter has a deep look at the background and historical features regarding the establishment of bioinformatics and computational biology and chemistry.

Section two comprises the single chapter (Chapter two) entitled: "Search Proteins Based on Human-Specific Availability Scores of Short Constituent Sequences: Identification of a WRWSH Protein in Human Testis". This chapter presents great information regarding in vitro-in silico studies.

Section three includes two chapters entitled: "Bioinformatics as a Tool for the Structural and Evolutionary Analysis of Proteins" and "Scaffolding Contigs Using Multiple Reference Genomes". These chapters offer brilliant information in association with the importance of bioinformatics and the related software tools and databases for analyzing proteins and genomes.

Section four contains two chapters, entitled: "Biological Evaluation and Molecular Docking Studies of a Double Active Pharmaceutical Ingredient, Benzalkonium Ibuprofenate" and "Hydrazones: An Important Scaffold to Construct Fluorescent Chemosensor for Biological purposes". These chapters provide fabulous information in the fields of computational biology and chemical monitoring.

Section five as the final section involves Chapter seven entitled: "Systems Glycobiology: Past, Present, Future". This chapter reveals the linkage between bioinformatics, databases, systems biology, and computational biology. An informative chapter with fantastic outcomes!

I, as the Editor of the book "Computational Biology and Chemistry", am honored and thankful to have marvelous cooperation and collaboration with valuable scientists from different countries and continents. They contributed as informative authors in this brilliant book. This book offers up-to-date information in the field to readers worldwide.

And finally, I have special thanks to my Italian colleague Dr. Nicola Bernabò from Università Degli Studi di Teramo, who collaborated with me as an invaluable Co-editor, Dolores Kuzelj the Author Service Manager, Lucija Tomicic-Dromgool and Martina Usljebrka Kauric the Commissioning Editors of IntechOpen for their excellent collaboration, management, and arrangement for preparing this valuable book.

Dr. Payam Behzadi
Department of Microbiology,
College of Basic Sciences,
Islamic Azad University,
Shahr-e-Qods Branch,
Tehran, Iran

Nicola Bernabò
Faculty of Bioscience,
University of Teramo,
Teramo, Italy

Section 1

Introduction

Chapter 1

Introductory Chapter: From Hard to Soft Biology

Payam Behzadi

1. Experimentation and computation

On the evening of Monday, April 13, 2020, during "the Wuhan-China virus (COVID-19) Home-Self Quarantine Era," I was drinking coffee and simultaneously searching on Google Scholar to find some valuable papers regarding "computational biology."

Among a mass of article links, an article entitled "laptop biology" [1] attracted me. I began to read this paper carefully and found some valuable terms including "hard science" and "soft science." Indeed, the term "soft science" was used for "experimentation, classification, observation and intuition," while the "hard science" term depicted "mathematics and algorithms" [1].

Then I checked https://www.thefreedictionary.com/ and searched for the terms "hard science" and "soft science." The results were as follows:

- "Hard Science: one of the natural or physical sciences, such as physics, chemistry, biology, geology, or astronomy, any of the natural or physical sciences, in which hypotheses are rigorously tested through observation and experimentation" (https://www.thefreedictionary.com/hard+science).

- "Soft Science: a science, such as sociology or anthropology, that deals with humans as its principle subject matter, and is therefore not generally considered to be based on rigorous experimentation, any of the scientific disciplines, as those which study human behavior or institutions, in which strictly measurable criteria are difficult to obtain" (https://www.thefreedictionary.com/Soft+science).

Although the meanings of these terms were very different from what I thought, I liked them. Due to this fact I found that it is better to use my terminology talent to represent new terms "hard biology" and "soft biology."

I used the term "hard biology" for experimental (in vitro, in vivo, and in situ investigations) biology. This term has a direct deal with experimentation and work bench within the wet labs [2].

In contrast to "hard biology," I used the term "soft biology" based on in silico or desktop work [2] and laptop biology [1].

I hope that these terms are useful for the readers of this book and other scientists around the world. With this background I begin the main text of the introductory chapter.

2. Biology and computer

Undoubtedly the famous physical chemist from the USA, Margaret Dayhoff (1925–1983), the mother and father of bioinformatics, was the key scientist who employed computers and the related software tools in biochemistry [3, 4].

Indeed, it was Dayhoff who understood the importance of computers and computational methods not only in biology but also in medicine [4].

In 1960, Dayhoff as the Associate Director of the National Biomedical Research Foundation began her collaboration with her physicist colleague, Robert S. Ledley, who, like Dayhoff, was interested in employing computers in biomedical sciences [3, 5, 6].

The outcome of their scientific collaboration (Dayhoff-Ledley) in the period of 1958–1962 led to a computer program COMPROTEIN (coded FORTRAN) which was designed for IBM 7090. The software COMPROTEIN (a de novo sequence assembler) was able to determine the primary structure of protein throughout the Edman peptide sequencing data [3, 7].

The amino acid one-letter coding system was founded by Dayhoff [3, 8]. Dayhoff and Eck continued their scientific activities by publishing the first edition of the invaluable book entitled *Atlas of Protein Sequence and Structure* in 1965 which involved 65 protein sequences [3, 4, 9]. The fourth edition of *Atlas of Protein Sequence and Structure* which was published in 1969 included more than 300 protein sequences. So, this atlas established the first database of biological sequence [3, 4]. Interestingly, the sequence alignment of biopolymers was started by proteins not DNA molecules. This claim is proven by representing a 12-sequenced DNA fragment in 1971 [4].

3. Molecular biology and computer

During the golden decade between the years 1970 and 1980, the DNA language was decoded. However, the genetic codes of 64 codons were decoded in 1968 [3, 10]. Sanger's sequencing method of DNA based on "plus and minus" strands was performed 25 years after the recognition of the first protein sequence [3, 11, 12]. 1979 is the historical year for using the first software for Sanger's DNA sequencing method. In the paper published by Rodger Staden in 1979 via the journal of *Nucleic Acids Research*, the applied programs including OVRLAP, XMATCH, and FILINS (coded FORTAN) were described [3, 13].

During the years of 1980–1990, the application of computational sciences significantly increased. In 1983, the polymerase chain reaction (PCR) was invented by Kary B. Mullis (1944–2019 (https://www.nytimes.com/2019/08/15/science/kary-b-mullis-dead.html)). Kary Banks Mullis as an American biochemist invented a valuable molecular method which was based on in vitro synthesis of DNA [14–16]. So, by the discovery of DNA molecules in the 1950s, and in consequence the early application of pro-computers, invention of molecular and sequencing methods, and utilizing Internet services within a short duration, it seems that several revolutionized features have happened in molecular biology [17].

Although computers and the related software tools were employed since the 1960s in biology, it was in the limited scales. I believe that by the invention of PCR as an in silico-in vitro (dry lab-wet lab) technology and its flying speed as a general molecular biology approach changed the traditional methodologies. By global generalization of PCR, the use of Internet services, computers, and software tools got significant acceleration. In this regard, designing different primers in large and global scales led to progression of in silico studies and appearance of dry labs within the wet labs.

Due to this fact, during a very short time, a mass of raw data was obtained by scientists around the world, and these data got stored within different biological databases like the National Center for Biotechnology Information (NCBI) (https://www.ncbi.nlm.nih.gov/), European Molecular Biology Laboratory (EMBL) (https://www.embl.org/), Kyoto Encyclopedia of Genes and Genomes (KEGG) (https://www.genome.jp/kegg/), and DNA Data Bank of Japan (DDBJ) (https://www.ddbj.nig.ac.jp/index-e.html) [18].

At the same time, these giant databases began to give more free software tools, information, and other services. These features have led to establishing 1637 free online databases (http://www.oxfordjournals.org/nar/database/c/) up to now [19].

Today, some databases including The Research Collaboratory for Structural Bioinformatics Protein Data Bank (RCSB PDB) (https://www.rcsb.org/) serve its global users for free. 3D macromolecular structure data is one of the most popular products which scientists and researchers use for free around the world [20].

Since 1971 when the US data center of RCSB PDB was founded, it provided digital data in biology for its users with open-access policy [20], an opportunity which all the consumers should be grateful for.

All in all, soft biology was founded in the 1960s with a low speed, but it accelerated by the invention of PCR in the 1980s. The PCR invention softened the science of biology throughout Internet services, an occurrence which resulted in progressive in silico studies, dry labs, and software tools.

Today, a biologist is recognized by her/his laptop, Internet connection, and filled USB flash drives with different bioinformatics software tools!

Hence, hard biology got softened by establishing the science of bioinformatics and is continued to be more softened by computational biology!

But the important question is:

"How does biology get more softened in the future?"

Conflict of interest

The authors declare no conflicts of interest.

Author details

Payam Behzadi
Department of Microbiology, College of Basic Sciences, Shahr-e-Qods Branch, Islamic Azad University, Tehran, Iran

*Address all correspondence to: behzadipayam@yahoo.com

References

[1] Hunter P. Laptop biology. EMBO Reports. 2005;**6**(3):208-210

[2] Penders B, Horstman K, Vos R. Walking the line between lab and computation: The "moist" zone. BioScience. 2008;**58**(8):747-755

[3] Gauthier J, Vincent AT, Charette SJ, Derome N. A brief history of bioinformatics. Briefings in Bioinformatics. 2019;**20**(6):1981-1996

[4] Moody G. Digital Code of Life: How Bioinformatics Is Revolutionizing Science, Medicine, and Business. London: John Wiley & Sons; 2004

[5] Ledley RS. Digital electronic computers in biomedical science. Science. 1959;**130**(3384):1225-1234

[6] November JA. Early biomedical computing and the roots of evidence-based medicine. IEEE Annals of the History of Computing. 2011;**33**(2):9-23

[7] Dayhoff MO, Ledley RS. Comprotein: A computer program to aid primary protein structure determination. In: Proceedings of the December 4-6, 1962, Fall Joint Computer Conference. New York, NY: ACM (Association for Computing Machinery); 1962

[8] International Union of Pure and Applied Chemistry (IUPAC)- the International Union of Biochemistry (IUB) Commission on biochemical nomenclature (CBN). A one-letter notation for amino acid sequences. Tentative rules. European Journal of Biochemistry. 1968;**5**:151-153

[9] Dayhoff MO. Atlas of Protein Sequence and Structure. National Biomedical Research Foundation. Washington, D.C.: Georgetown University Medical Center; 1972

[10] Crick FH. The origin of the genetic code. Journal of Molecular Biology. 1968;**38**(3):367-379

[11] Sanger F, Thompson E. The amino-acid sequence in the glycyl chain of insulin. 1. The identification of lower peptides from partial hydrolysates. Biochemical Journal. 1953;**53**(3):353

[12] Sanger F, Thompson E. The amino-acid sequence in the glycyl chain of insulin. 2. The investigation of peptides from enzymic hydrolysates. Biochemical Journal. 1953;**53**(3):366

[13] Staden R. A strategy of DNA sequencing employing computer programs. Nucleic Acids Research. 1979;**6**(7):2601-2610

[14] Kadri K. Polymerase Chain Reaction (PCR): Principle and Applications. Perspectives on Polymerase Chain Reaction. Croatia: IntechOpen; 2019

[15] Mullis KB, Faloona FA. Specific Synthesis of DNA in Vitro Via a Polymerase-Catalyzed Chain Reaction. Recombinant DNA Methodology. San Diego: Academic Press, Elsevier; 1989. pp. 189-204

[16] Pai-Dhungat J. Kary Mullis—Inventor of PCR. Journal of the Association of Physicians of India. 2019;**67**:96

[17] Bartlett JM, Stirling D. A Short History of the Polymerase Chain Reaction. PCR Protocols. Totowa New Jersy: Humana Press, Springer; 2003. pp. 3-6

[18] Zou D, Ma L, Yu J, Zhang Z. Biological databases for human research. Genomics, Proteomics & Bioinformatics. 2015;**13**(1):55-63

[19] Rigden DJ, Fernández XM. The 27th annual Nucleic Acids Research database

issue and molecular biology database collection. Nucleic Acids Research. 2020;**48**(D1):D1-D8

[20] Burley SK, Berman HM, Bhikadiya C, Bi C, Chen L, Di Costanzo L, et al. RCSB Protein Data Bank: Biological macromolecular structures enabling research and education in fundamental biology, biomedicine, biotechnology and energy. Nucleic Acids Research. 2019;**47**(D1):D464-DD74

Section 2

Wet Labs and Dry Labs: From In Vitro to In Silico Studies

Chapter 2

Search for Human-Specific Proteins Based on Availability Scores of Short Constituent Sequences: Identification of a WRWSH Protein in Human Testis

Shiho Endo, Kenta Motomura, Masakazu Tsuhako, Yuki Kakazu, Morikazu Nakamura and Joji M. Otaki

Abstract

Little is known about protein sequences unique in humans. Here, we performed alignment-free sequence comparisons based on the availability (frequency bias) of short constituent amino acid (aa) sequences (SCSs) in proteins to search for human-specific proteins. Focusing on 5-aa SCSs (pentats), exhaustive comparisons of availability scores among the human proteome and other nine mammalian proteomes in the nonredundant (nr) database identified a candidate protein containing WRWSH, here called FAM75, as human-specific. Examination of various human genome sequences revealed that FAM75 had genomic DNA sequences for either WRWSH or WRWSR due to a single nucleotide polymorphism (SNP). FAM75 and its related protein FAM205A were found to be produced through alternative splicing. The FAM75 transcript was found only in humans, but the FAM205A transcript was also present in other mammals. In humans, both FAM75 and FAM205A were expressed specifically in testis at the mRNA level, and they were immunohistochemically located in cells in seminiferous ducts and in acrosomes in spermatids at the protein level, suggesting their possible function in sperm development and fertilization. This study highlights a practical application of SCS-based methods for protein searches and suggests possible contributions of SNP variants and alternative splicing of FAM75 to human evolution.

Keywords: availability score, short constituent sequence (SCS), alternative splicing, single nucleotide polymorphism (SNP), testis, FAM75, FAM205A, human genome, human proteome

1. Introduction

The human species has unique traits among animals. It is well known that morphological and physiological traits such as erect bipedalism, speech and language, and long reproductive period are very different from those of other primate species. Only humans have high intelligence that fosters sophisticated communications and complex societies. This intelligence is related to continuous brain development after birth in humans, which is not observed in other great apes, including chimpanzees [1]. The evolutionary emergence of these unique traits in humans likely contributes to human speciation. The simplest hypothesis to explain human uniqueness is that it originates from the uniqueness of constituent molecules (i.e., genes and proteins) themselves. In this "constituent hypothesis," humans have unique genes and proteins that do not exist in chimpanzees. A contrasting hypothesis is that constituent molecules are similar between humans and chimpanzees, but they are regulated differently in these species. That is, in this "regulatory hypothesis," a similar set of proteins may be produced but at different times (heterochrony), in different locations (heterotopy), in different amounts (heterometry), and in different usage (heterotypy) [2]. These regulatory changes in gene expression seem to be evolutionarily parsimonious and, indeed, are supported by comparative observations at phenotypic levels [3].

One line of support for the regulatory hypothesis comes from genomics and developmental expression studies. Following the announcement of a human genome release [4], the genomes of other great apes were sequenced [5–7]. Comparisons of DNA sequences between humans and chimpanzees have revealed that nucleotide differences are only 1.23% in aligned sequences, and most of these differences are thought to be functionally insignificant [5]. Further rigorous comparisons throughout these genomes have revealed that nucleotide differences are 4% and that they are mostly located in noncoding regions [8]. The expression patterns of some genes are different between humans and chimpanzees during development [9–12]. Differences in transcriptomes have revealed that species differences in expression patterns are tissue-dependent and that testes have the greatest difference [13, 14]. It has been speculated that the accumulation of small expression or regulatory differences leads to large phenotypic differences between humans and chimpanzees [14]. On the other hand, while these findings support the regulatory hypothesis, they do not necessarily reject the constituent hypothesis [15, 16]. RNA-mediated mechanisms for novel genes have been proposed together with the "out of the testis" hypothesis, in which testis is considered a tissue for experimenting with new genes [16]. Comparisons among transcriptomes in primates have revealed that many genes for spermatogenesis in testes, which likely inhibit apoptosis when mutated, are positively selected [17, 18].

Although these genome comparison studies advance this field, there are a few inherent problems. First, their results are heavily dependent on database quality because of their methodological nature. Most genome sequences were draft sequences at the time of public release, likely containing numerous sequencing and assembling mistakes. For example, the previous chimpanzee genome was assembled in reference to the human genome, which means that genomic regions in chimpanzees that are different from those in the human genome may have been assembled to create false sequences, although continuous revisions have been made [19]. Even in the human genome, many previous gene records generated by automated assemblers have been removed after revisions. Moreover, population sampling bias from the sequenced genome cannot be avoided when samples from a small number of individuals are sequenced. The case of a transcription factor, FOXP2 (forkhead box P2), is an object lesson: FOXP2 has been proposed to have played a key role in human-specific evolution by assisting speech and language [20], but that evidence is likely to be weak and probably incorrect because of sampling bias [21].

Second, such genome comparisons are largely based on sequence alignments [22, 23]. Although sequence alignment methods are powerful and probably the most important in comparison studies, sequences that do not contain relatively long regions of similarity cannot be compared well. In other words, short sequences that do not extend to longer similarities are discarded as noise [22]. Although this strategy is highly successful, it assumes that nonaligned short sequences are not important, which may not always be true. There may still be important differences undiscovered where alignments are not possible.

An approach to the second issue above is to develop alignment-free methods. The advantage of the alignment-free approach is that any collections of proteins can be compared quantitatively. Although various types of alignment-free approaches have been developed [24, 25], including our previous attempts to use membrane topology [26] and a self-organizing map [27], the alignment-free approach in the present study is based on the "availability" (frequency bias) of short constituent sequences (SCSs) of amino acids (aa) in proteins [28–33]. The length of SCSs can be 2 aa (doublet), 3 aa (triplet), 4 aa (quartet), 5 aa (pentat), and more in a given protein. This SCS-based analysis is basically similar to other related analyses for amino acid sequence patterns that were called under different terms with slightly different mathematical operations: oligopeptide patterns [34–39], amino acid sequence repertoire [40], peptide vocabulary [41], *n*-gram [42, 43], *n*-tuple [44], and pseudo amino acid composition [45–47]. There are some noteworthy recent studies that encourage this line of approach: for example, nonrandom distributions of 5-aa SCS are demonstrated in the current proteome databases [38], confirming the previous finding that biological bias occurs in protein coding [28, 29]. Among these existing studies, our approach is operationally one of the simplest, and it emphasizes analogies between languages and protein sequences [32, 33]. Encouragingly, linguistic aspects of proteins have been noted in other studies [48, 49].

In our approach, protein sequences are considered to be composed of many SCSs. Importantly, the number of possible SCSs is limited because a protein is composed of just 20 kinds of amino acids; there are 400 ($=20^2$) permutations of 2-aa SCSs (doublets), 8000 ($=20^3$) permutations of 3-aa SCSs (triplets), 160,000 ($=20^4$) permutations of 4-aa SCSs (quartets), and 3,200,000 ($=20^5$) permutations of 5-aa SCSs (pentats). Frequencies of individual SCSs in a given protein database can be inferred theoretically based on frequencies of component amino acids, which is called the expected frequency (E). On the other hand, real frequencies of individual SCSs (R) in a given protein database can be obtained through database searches. The availability score (A) of a given SCS in a protein database can be simply defined as $A = (R - E)/E$. Availability scores thus indicate biological frequency bias that might have occurred for functional or historical reasons during protein evolution. In other words, availability scores (A) of SCSs are used instead of simple real frequencies (R) of SCSs to exclude noise from random occurrence.

Among *n*-SCSs, we state that 5-aa SCSs (pentats) are optimal for analyses for the following reasons [28, 29, 33]. First, they are practically manageable in number (exactly 3,200,000 different pentats) in our computational system. Higher computational power, which is sometimes not practical, is required to use 6-aa or longer SCSs. Second, the number of possible SCSs should be reasonably comparable to or smaller than the number of existing SCSs in a biological database. The use of 6-aa or longer SCSs would result in many nonexistent SCSs in the database because the number of possible 6-aa (or longer) SCSs is much larger than the number of existing SCSs in a given database. Third, 5-aa SCSs are likely structurally reasonable units (or "blocks") to build functional protein structures [50–54]. Fourth, it was suggested that small stretches of proteins are often recognized in protein interactions. For example, T-cell receptors recognize 5-aa SCSs as antigens in the process of antigen presentation, and this fact relates to the frequency bias of SCSs in parasites to avoid recognition by the T-cell receptors of the host [41]. Specificities of immune responses are thus likely influenced by SCSs in expressed proteins in a given organism, as also suggested by usage of rare SCSs as immune adjuvant vaccines [39].

Furthermore, rare SCS sequences evolved as untranslatable sequences in bacteria as a mean of translational control [40].

Using this simple concept of availability score, secondary structure characterization has been performed; SCS frequencies (and thus availability scores) are different among different secondary structures [30]. Availability scores are also different between parallel and antiparallel β-strands [31]. This approach is also relevant to identifying sequence motifs in some, although not all, proteins [32]. It has been shown that triplet compositions in proteomes may reflect phylogenetic relationships [32, 37, 53]. We believe that this approach is applicable to understanding species specificity.

We have implemented several applications as the SCS Package that informatically examine protein sequences [33]. Among them, we have built an application for identifying species-specific SCSs. In the present study, we compared human and other 9 mammalian proteomes based on availability analysis of 5-aa SCSs (pentats) to identify human-specific pentats. We hypothesized that a protein containing the identified human-specific pentat would be unique to humans and might have played a role in human evolution.

2. Materials and methods

2.1 The SCS package

Assuming that small changes in amino acids in proteins (or corresponding nucleotide changes in DNA) contribute significantly to phenotypic differences between humans and chimpanzees, the concept of SCS-based methods is to detect small amino acid usage differences between species in an alignment-independent manner. The SCS package is an open web service containing six applications (plus the latest application to analyze idiom networks under development [55]) for protein analyses (http://scspackage.ads.ie.u-ryukyu.ac.jp/) [33]. These applications run in reference to the database downloaded from the nonredundant (nr) database of the NCBI (National Center for Biotechnology Information, U.S. National Library of Medicine, Bethesda), which was downloaded on August 20, 2015. Because T-cell receptors, B-cell receptors, and antibodies are produced by somatic recombination and hypermutation, protein records containing the following keywords in sequence names were excluded: anti, IgG, IgM, IgA, IgD, IgE, BCR, TCR, B-cell receptor, T-cell receptor, Ig, and immunoglobulin. A complete match, including spaces, was required to be excluded. Frequencies and availability scores for all possible n-aa SCSs (n = 3, 4, and 5 in the current SCS package) in the database were calculated and stored in the database [56]. For species comparison, each record in the downloaded database was sorted into its original species to produce species-specific proteome databases.

In this study, we focused on 5-aa SCSs (pentats). For multiple species comparison, the availability score difference, ΔA, was calculated; for example, when a human 5-aa SCS had an availability score of 10 and the availability scores of that SCS in gorilla, pig, and mouse were 5, 3, and 2, respectively, the human ΔA was calculated as follows: $\Delta A = 10 \times 3 - (5 + 3 + 2) = 20$, where the multiplicative factor ($\times 3$) comes from the number of species to be compared. In this way, ΔA scores were assigned to all 3,200,000 pentats. The following nine species were used to obtain ΔA for human (*Homo sapiens*): chimpanzee (*Pan troglodytes*), gorilla (*Gorilla gorilla*), orangutan (*Pongo abelii*), mouse (*Mus musculus*), rat (*Rattus norvegicus*), opossum (*Monodelphis domestica*), platypus (*Ornithorhynchus anatinus*), cow (*Bos taurus*), and pig (*Sus scrofa*).

2.2 Bioinformatics web services

After identifying FAM75 using the SCS package, available information on FAM75 and FAM205A was gathered using various web sites. The location FAM75/FAM205A on chromosomes and their single nucleotide polymorphism (SNP) variants were searched using Map Viewer (www.ncbi.nlm.gov/mapview/) in the NCBI server. For various information on human transcripts, we referred to H-InvDB (www.h-invitational.jp/hinv/ahg-db/index_ja.jsp) [57]. This site provides curated information on gene structure, splicing variants, functional RNAs, protein functions, functional domains, intracellular distribution, metabolic pathways, three-dimensional structures, disease relationships, genetic polymorphism (SNPs, indels, microsatellites, and others), gene expression profiles, molecular evolutionary characters, protein–protein interactions, and gene families. Tissue-specific expression profiles were searched using H-ANGEL (http://www.h-invitational.jp/hinv/h-angel/wge_top.cgi?), a database for human gene expression profiles, in the H-InvDB server. Information on alternative splicing variants of the nonhuman primates and mouse was obtained from Map Viewer in NCBI. We referred to the following latest annotations: chimpanzee (Annotation Release 103), western gorilla (Annotation Release 100), Sumatran orangutan (Annotation Release 102), and laboratory mouse (Annotation Release 106). We frequently used protein BLAST in the NCBI server for conventional similarity search (blast.ncbi.nlm.nih.gov/Blast.cgi?PAGE=Proteins) [22] and performed multiple sequence alignments when necessary using MEGA7 (www.megasoftware.net/) [58]. In addition, the cDNA sequence of FAM75 was subjected to RegRNA analysis (regrna.mbc.nctu.edu.tw/html/about.html) to identify any possible sequence motifs in FAM75 mRNA [59].

To further examine SNP variants in human populations, we used dbSNP (www.ncbi.nlm.nih.gov/snp/) [60] and the 1000 Genome Project by IGSR (The International Genome Sample Resource) (www.internationalgenome.org) [61]. Protein expression was examined using The Human Protein Atlas (www.proteinatlas.org) [62, 63]. This site contains immunohistochemical data for various human tissues. For identification of transmembrane domains in FAM75 and FAM205A, SOSUI (harrier.nagahama-i-bio.ac.jp/sosui/) [64] and TMHMM (www.cbs.dtu.dk/services/TMHMM/) [65] were used. For the subcellular distributions of FAM75 and FAM205A, PSORT II Prediction (psort.hgc.jp/form2.html) [66] was used. Pfam (pfam.xfam.org) [67] was used for the identification of protein families. Two applications of the SCS Package, "sequence analysis based on availability scores of short constituent amino acid sequences" (scspackage.ads.ie.u-ryukyu.ac.jp/sequence-analysis.php) [32, 33] and "extraction of idiomatic connections between triplets in proteins" (scspackage.ads.ie.u-ryukyu.ac.jp/extraction-of--idiomatic-connections.php) [33], were used to identify possible functional sites in FAM75. EMBOSS Pepwindow (emboss.sourceforge.net/index.html) [68] was used for the Kyte-Doolittle hydropathy plot. These web sites were accessed mainly in 2017 and 2018 and were reconfirmed in 2019.

2.3 Human cDNA samples for tissue expression profiling

For the cDNA template, we purchased human MTC (multiple tissue cDNA) panels I and II (Takara Bio, Kusatsu, Shiga, Japan). The panels contain first-strand cDNA from polyA$^+$ RNA and are free from genomic DNA. The amounts of cDNA are approximately 1.0 ng/μL and are normalized to four housekeeping genes, phospholipase A2, G3PDH (glyceraldehyde-3-phosphate dehydrogenase), β-actin, and α-tubulin, which makes it possible to compare expression levels among different tissues. Panels I and II together contain cDNA samples from the following 16 human tissues: heart, brain, placenta, lung, liver, skeletal muscle, kidney, pancreas, spleen, thymus, prostate, testis, ovary, small intestine without mucosal lining, colon with mucosal lining, and peripheral blood leukocyte. Each tissue sample was pooled from 1 to 550 Caucasians, and the testis sample was pooled from 45 Caucasians aged 14–64, according to the manufacturer's specifications.

2.4 PCR primers

Based on the cDNA sequence of FAM75, we designed two sets of PCR primers for nested PCR using Primer-BLAST (www.ncbi.nlm.nih.gov/tools/primer-blast/). The first set was to amplify both FAM75 and FAM205A from the consensus region, and the second set was to amplify FAM205A from the region that is present only in FAM205A. For the first set, the first-round forward primer was 5'-TTACCAGG-TACTGTCACTGAACAC-3′, and its paired reverse primer was 5'-TTCTGAAGC-TAGACTCTGTAAGGC-3′. This first round of PCR was expected to amplify 1387 bp. The second-round (nested) forward primer was 5'-AGTTGTACA-GACGTTGCAAAAGAG-3′, and its paired reverse primer was 5'-TTTCTGAAGC-TAGACTCTGTAAGGC-3′. This second round (nested) PCR was expected to amplify 1097 bp.

For the second set, the first-round forward primer was 5'-ATATCCCTTATA-CATCTATGGCTCCATCTTC-3′, and its paired reverse primer was 5'-TTTTATTTCTGAAGCTAGACTCTGTAAGGC-3′. This round of PCR was expected to amplify 3608 bp. The second-round (nested) forward primer was 5'-GTATGTCTTTAGATCAGAGTCTGGAGTTTC-3′, and its paired reverse primer was 5'-TTTATTTCTGAAGCTAGACTCTGTAAGGCTG-3′. This round of PCR was expected to amplify 3206 bp.

2.5 PCR conditions

We used an Astec PC320 thermal cycler (Fukuoka, Japan) and Tks Gflex DNA polymerase (Takara Bio) for PCR. According to the manufacturer's specifications, this DNA polymerase has high fidelity; the error rate was reported to be 0.0131%.

The original cDNA sample from the human MTC panels (Takara Bio) was diluted 10 times to make PCR template samples. The following solutions were mixed to start PCR: Gflex PCR buffer 12.5 μL, deionized water 8.5 μL, DNA polymerase 0.5 μL, forward primer 0.5 μL, reverse primer 0.5 μL, and cDNA template 2.5 μL in a total amount of 25.0 μL. The nested PCR was performed using 2.5 μL reaction solution from the first-round PCR. In both the first and second (nested) rounds, a negative control was performed using deionized water without template cDNA.

The first PCR cycles were performed as follows: an initial denaturing step at 94°C (5 min); 10 cycles of 98°C (30 s), 60°C (30 s; −0.5°C/cycle), and 68°C (1 min); 30 cycles of 98°C (30 s), 55°C (30 s), and 68°C (1 min); and a last extension step at 68°C (30 s). The second (nested) PCR cycles were the same as the first PCR cycles except the duration of the initial denaturing step at 94°C (1 min). In both the first and second PCRs, the first 10 cycles were subjected to stepwise temperature reduction (i.e., touch-down PCR); the first cycle was 60.0°C, the second cycle was 59.5°C, and the third cycle was 59.0°C, and so on.

Positive controls were performed with G3PDH primers that were supplied in the Human MTC Panels (Takara Bio) from the manufacturer. The PCR product was expected to be 938 bp. The primer sequences were as follows: 5'-TGAAGGTCG-GAGTCAACGGATTTGGT-3′ for the forward primer and 5'-CATGTGGGCCAT-GAGGTCCACCAC-3′ for the paired reverse primer. PCR cycles were as follows: an initial denaturing step at 95°C (1 min); 38 cycles of 95°C (30 s) and 68°C (3 min); and a final extension step at 68°C (3 min).

PCR products (1.0 μL) were subjected to 0.8% agarose gel electrophoresis in TAE buffer and stained with ethidium bromide for visualization. The PCR products were run with λHindIII DNA size marker (New England Biolabs, Ipswich, MA, USA).

3. Results

3.1 Identifying candidate human-specific pentats

Availability scores (A) were given to all possible pentats in the human proteome database. Among them, the top 10 pentats with the highest availability scores were HHHHH (rank 1; A = 1837), MYGCD (rank 2; A = 1770), MRYFY (rank 3; A = 1321), WYWHF (rank 4; A = 1262), PEYWD (rank 5; A = 1140), MYQWW (rank 6; A = 1100), HSMRY (rank 7; A = 1096), NWHWA (rank 8; A = 1041), WWNFG (rank 9; A = 1007), and AWWNF (rank 10; A = 928). Similarly, availability scores were given to all possible pentats in nine other mammalian proteome databases. Using the "extraction of species-specific amino acid sequences" program in the SCS Package, availability difference scores (ΔA) were calculated for the human proteome. When pentats were ranked according to ΔA for humans, the top 10 pentats with the highest scores were MYGCD (rank 1; ΔA = 15,180), MRYFY (rank 2; ΔA = 11,777), WYWHF (rank 3; ΔA = 10,683), PEYWD (rank 4; ΔA = 9961), HSMRY (rank 5; ΔA = 9695), MYQWW (rank 6; ΔA = 9377), NWHWA (rank 7; ΔA = 9337), GQWRW (rank 8; ΔA = 8255), AWWNF (rank 9; ΔA = 7939), and EYWDR (rank 10; ΔA = 7878), showing similar but different rank orders from the human proteome alone. These pentats had large ΔA values, indicating that they are strongly preferred in human proteins.

Among the ΔA rank order of pentats, we focused on pentats that showed the lowest possible availability scores (A = −1) in all other nine mammalian proteome databases, meaning that these pentats did not exist in the nonhuman proteomes at all. We found WRWSH at rank 204 (ΔA = 1720) and MMFGC at rank 226 (ΔA = 1594) that met this criterion. However, MMFGC was found to be a false-positive, because this pentat was located exclusively in immunological proteins that could be subject to somatic recombination and hypermutation. Therefore, we decided to focus on WRWSH hereafter.

3.2 Human proteins containing WRWSH

Human proteins containing WRWSH were identified using the "search for amino acid sequences of species" program, one of the SCS Package programs. Among all 148 hits, 16 hits were related to "mucin-19-like isoform," 55 hits to "glycine-rich cell wall structural protein," 28 hits to "RNA-binding protein," 48 hits to "uncharacterized transmembrane protein," and 1 hit to "unnamed protein product." Unfortunately, these sequences except the last one, "unnamed protein product," were all "predicted" informatically as parts of "*Homo sapiens* Annotation Release 106" [69], and they were all removed from the latest annotation, "*Homo sapiens* Annotation Release 109" [70]. Because their status was uncertain at this point (although they resembled real protein sequences with a long open reading frame), they were not pursued in the present study. On the other hand, "unnamed protein product [*Homo sapiens*] (Accession No. BAC86357.1)", here called "FAM75" based on the name of putative domain that it contained, was validated in the latest annotation [70], and thus, we pursued this protein for further investigation.

3.3 FAM75 and its related FAM205A

According to the NCBI record, FAM75 is a protein containing 1014 aa, and its cDNA coding sequence was 3274 bp (Accession No. AK125949.1). It is important to stress that FAM75 has been identified as cDNA from NEDO human cDNA sequencing project (www.nite.go.jp/en/nbrc/genome/project/annotation/cdna.html), and thus this protein is not likely an error product from genome sequencing. A protein BLAST search using FAM75 as a query identified the record "protein FAM205A [*Homo sapiens*] (Accession No. NP_001135389.1)." This protein record was closely related to the mRNA record "*Homo sapiens* family with sequence similarity 205 member A (FAM205A), mRNA (Accession No. NM_001141917.1)." The BLAST result showed that the identity score was 99%; 1003 aa were identical among 1014 aa. The record showed that FAM205A contained 1335 aa, and its cDNA coding sequence was 4311 bp. Thus, it was longer than FAM75. A DNA sequence comparison between FAM75 and FAM205A revealed that 16 bases were different (**Table 1**). When the FAM205A genomic DNA sequence (Accession No. NG_052658.1) was compared with its cDNA sequence, these 16 bases were identical (**Table 1**). Between FAM75 and FAM205A, 11 amino acids were different.

FAM75 BAC86357.1, AK125949.1 (cDNA)		FAM205A NG_052658.1 (genomic DNA)		FAM205A NP_001135389.1, NM_001141917.1 (cDNA)	
122	T(S)	8332	C(P)	1144	C(P)
139	C(H)	8349	T(H)	1161	T(H)
144	C(S)	8354	T(E)	1166	T(E)
584	G(V)	8794	A(M)	1606	A(M)
888	C(S)	9098	T(L)	1910	T(L)
894	A(H)	9104	G(R)	1916	G(R)
958	C(G)	9168	T(G)	1980	T(G)
1279	A(P)	9489	G(P)	2301	G(P)
1530	T(V)	9740	A(E)	2552	A(E)
1540	T(P)	9750	C(P)	2562	C(P)
2085	T(I)	10295	G(S)	3107	G(S)
2226	A(E)	10436	G(G)	3248	G(G)
2267	T(Y)	10477	C(H)	3289	C(H)
2391	A(H)	10601	G(R)	3413	G(R)
2582	T(S)	10792	G(A)	3604	G(A)
3145	T	11355	C	4167	C

Note: Numbers in this table indicate those of bases in the original records. Corresponding amino acids are shown in parenthesis. The shaded bases (and corresponding amino acids) correspond to the candidate human-specific pentat WRWSH in FAM75.

Table 1.
Different DNA bases and protein amino acids between FAM75 (unnamed protein product) and FAM205A in the NCBI records.

Interestingly, FAM205A in that record had WRWSR instead of WRWSH; the DNA sequences corresponding to the last amino acid of WRWS (H/R) were A (adenine) in FAM75 cDNA and G (guanine) in FAM205A cDNA and gDNA. These results suggest that the two products are closely related and may be produced from the same genomic site by alternative RNA splicing.

3.4 Gene structures: alternative splicing and polymorphism

A UniGene search revealed that the FAM7/FAM205A gene was located at 9p13.3 on chromosome 9 in the human genome [71]. As expected, their exon-intron structures were different (**Figure 1**). FAM75 had a single exon, whereas FAM205A had four exons. The exon of FAM75 had high homology with the fourth exon of FAM205A. The 5'-UTR of FAM75 also corresponded to the fourth exon of FAM205A. Clearly, these two RNA transcripts and their proteins are products of alternative splicing from the same genomic locus.

H-InvDB revealed two additional splicing variants (HIT000496944 and HIT000496575) from the same locus at 9p13.3 (**Figure 1**). The record HIT000496944 in the NCBI database was "*Homo sapiens* cDNA FLJ51393 complete code (AK302320.1)," and the record HIT000496575 was "*Homo sapiens* cDNA FLJ58301 complete code (AK301951.1)," both named "unnamed protein product." These are splicing variants, but among them, only FAM75 lacked the first 255-bp exon, indicating that the translation initiation sites are different in these mRNAs.

Because not all RNA transcripts are translated into proteins, we used a RegRNA search of UTRs (untranslated regions) to examine the integrity of the FAM75 mRNA. The RegRNA search revealed that the 5'-UTR of FAM75 had an internal ribosome entry site (IRES) [72–74] among other motifs, suggesting that the FAM75 mRNA is likely translated into proteins.

3.5 WRWSH and WRWSR in human populations

The G/A difference in FAM75/FAM205A in genomic DNA (corresponding to the H/R difference in WRWSH or WRWSR) was confirmed to be a SNP in humans, according to dbSNP. We found that this SNP was widespread in the human genome, and the G/A ratio was dependent on regional populations, as revealed by the 1000 Genomes Project (**Figure 2**). Among human populations, African populations had a high G frequency (i.e., WRWSR); the three highest G-frequency populations were Gambian in Western Division (96.16%); Yoruba in Ibadan,

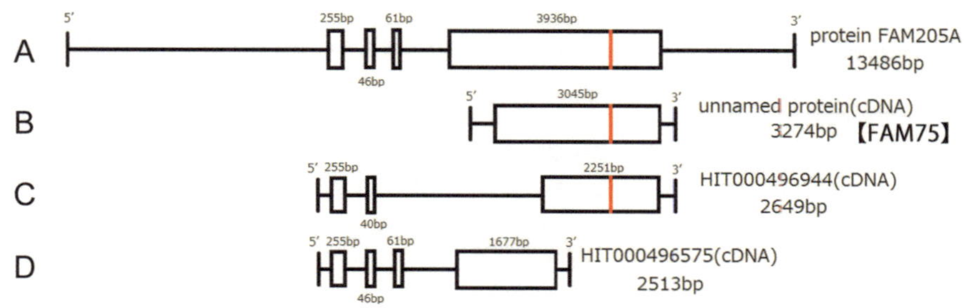

Figure 1.
mRNA structures of FAM75, FAM205A, and their related transcripts from the same genomic locus in the human genome. (A) FAM205A from UniGene. (B) Unnamed protein (FAM75) from UniGene. (C) HIT000496944 from H-InvDB. (D) HIT000496575 from H-InvDB.

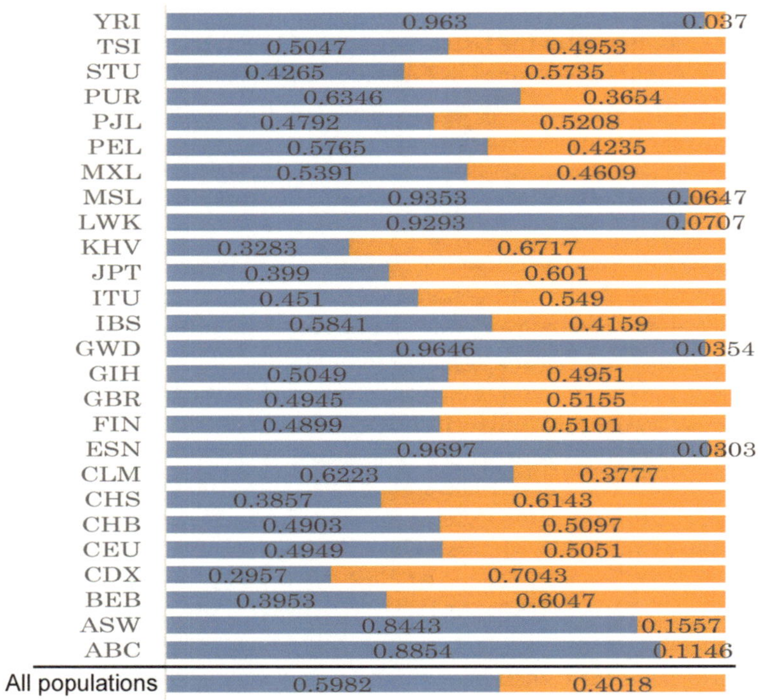

Figure 2.

Genomic G/A ratio at the SNP site of the candidate human-specific pentat WRWSH in FAM75/FAM205A in various human populations. Abbreviations of populations or samples: YRI (Yoruba in Ibadan, Nigeria), TSI (Toscani, Italy), STU (Sri Lankan Tamil, UK), PUR (Puerto Rican, Puerto Rico), PJL (Punjabi in Lahore, Pakistan), PEL (Peruvian in Lima, Peru), MXL (Maxican ancestry in Los Angeles, CA, USA), MSL (Mende, Sierra Leone), LWK (Luhya in Webuye, Kenya), KHV (Kinh in Ho Chi Minh City, Vietnam), JPT (Japanese in Tokyo, Japan), ITU (Indian Telugu, UK), IBS (Iberian populations, Spain), GWD (Gujarati Indians in Houston, TX, USA), GBR (British from England and Scotland), FIN (Finnish, Finland), ESN (Esan, Nigeria), CLM (Colombian in Medellín, Colombia), CHS (Han Chinese south, China), CHB (Han Chinese in Beijing, China), CEU (Utah residents (CEPH) with northern and Western European ancestry), CDX (Chinese Dai in Xishuangbanna), BEB (Bengali, Bangladesh), ASW (African ancestry in SW, USA), and ABC (African Caribbean, Barbados).

Nigeria (96.30%); and Mende in Sierra Leone (93.53%). In contrast, Asian and European populations had relatively high A frequency (i.e., WRWSH); the three highest A-frequency populations were Chinese Dai in Xishuangbanna (70.43%); Kinh in Ho Chi Minh City, Vietnam (67.17%); and Han Chinese South, China (61.43%).

3.6 Homologous proteins and alternative splicing products in other animals

Here, we searched for homologous proteins for FAM75 and FAM205A in other animals. Among the nine mammals used for the initial identification for WRWSH, homologous proteins for FAM205A were identified by BLAST search (**Table 2**); all proteins were FAM205A homologs, and all proteins in primates (chimpanzee, gorilla, and orangutan) contained WRWSR (corresponding to FAM205A) but not WRWSH (corresponding to FAM75), suggesting that WRWSH in FAM75 may be unique in humans among primates. In other nonprimate animals that were examined here, this pentat sequence was either not conserved at all or nonexistent.

To further examine whether splicing variants exist in other great apes, we checked the genome loci and transcript data using Map Viewer. In chimpanzees (**Figure 3**) and gorillas (not shown), there were no alternative splicing transcripts from this locus. In orangutans (not shown), there were three isoforms, the X1, X2,

	Name in FASTA file	ID	Pentat
Human	unnamed protein	BAC86357.1	WRWSH
Human	protein FAM205A	NG_052658.1	WRWSR
Chimpanzee	protein FAM205A	XP001164235.2	WRWSR
Gorilla	protein FAM205A like	XP_004048025.1	WRWSR
Orangutan	protein FAM205A isoform	XP_009242592.1	WRWSR
Mouse	predicted gene 12,429 isoform	XP_011248363.1	SLQAQ
Rat	protein FAM205-A isoform	XP_008774156.1	SQQGH
Opossum	protein FAM205-A like isoform	XP_007498908.1	HVGNR
Platypus	protein FAM205A	XP_007657228.1	:::::
Cow	protein FAM205A	XP_001253501.1	WQRRH
Pig	—	—	—

Note: Amino acid sequences are conceptual translation from genomic data. Red letters indicate amino acids different from those of the human pentat WRWSH. No corresponding pentat was found in the platypus (:::::), and no homologous protein was found in the pig (—).

Table 2.
Amino acid pentat sequences from mammals that are homologous to the human WRWSH in FAM75.

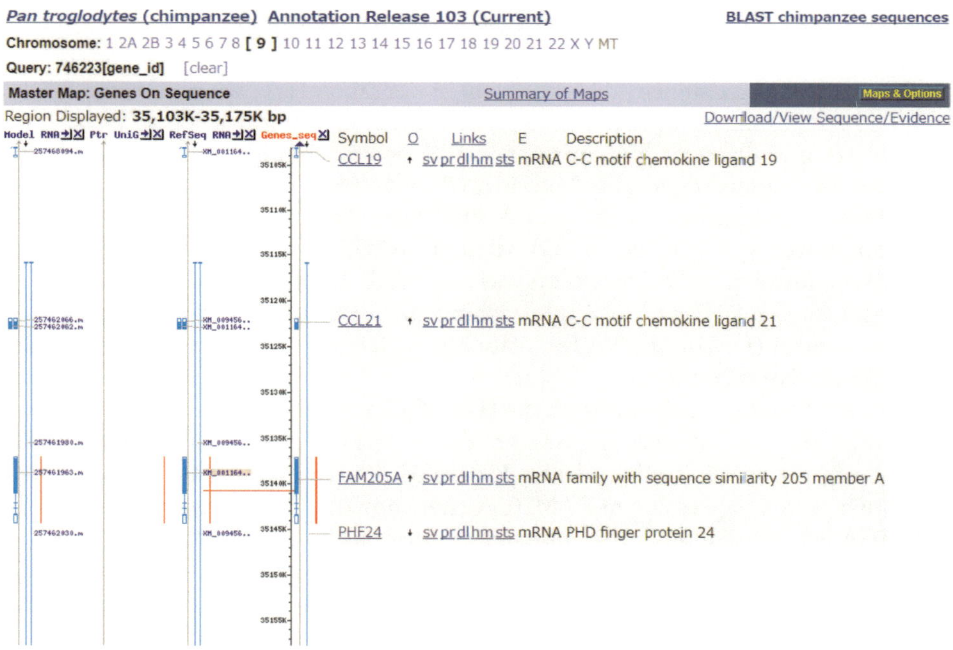

Figure 3.
FAM205A and its surrounding locus of chromosome 9 in the chimpanzee genome.

and X3 transcripts, from this locus. However, these transcripts were very similar to one another, and they were all considered FAM205A homologs containing WRWSR. We also examined the genome of the mouse as a representative nonprimate mammal (not shown). There were three transcript variants: "predicted gene 12429 isoform X1, X2" and "predicted gene 12429." They all contained SLQAQ instead of WRWSH in these proteins, and their splicing patterns were different from those of FAM75. We confirmed that human splicing patterns (**Figure 4**) were

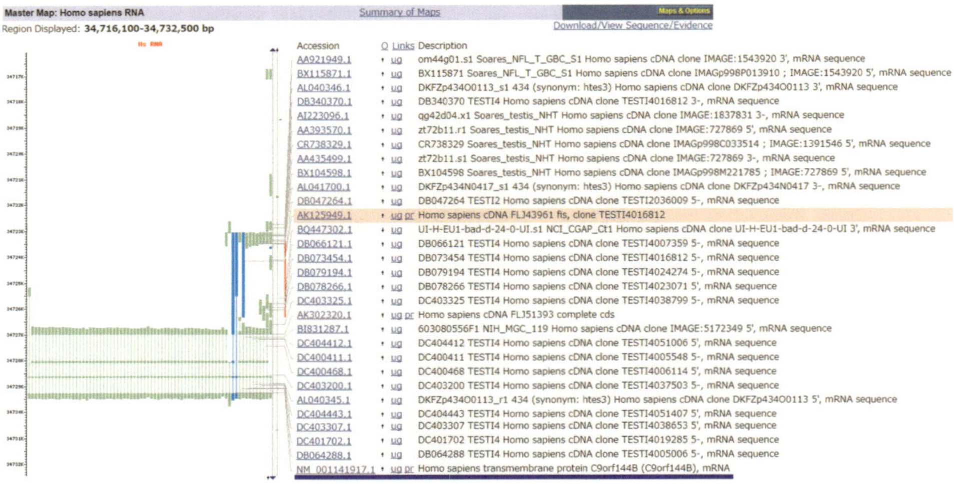

Figure 4.
FAM205A/FAM75 and its surrounding locus of chromosome 9 in the human genome. FAM75 is highlighted in pink, and FAM205A is underlined in blue.

different from those of these mammals. Therefore, we conclude that the FAM75 transcript was found only in humans.

3.7 Testis-specific expression of FAM75 and FAM205A

To examine its existence and expression in our laboratory, we performed RT-PCR (reverse transcription polymerase chain reaction) using two sets of PCR primers using 16 different human-tissue cDNA pools as templates. The first set of primers was designed to amplify both FAM75 and FAM205A (**Figure 5A**), and the second set was designed to amplify FAM205A only (**Figure 5B**). Due to their overlapping nature, exclusive amplification of FAM75 was not possible. In both primer sets, testis-specific expression was observed. A positive control using a primer set for G3PDH showed amplification from all tissues (**Figure 5C**), and a negative control (without cDNA template but with experimental primer sets) did not show any amplification.

Our results were consistent with the H-ANGEL expression database; in this database, FAM75 and FAM205A were not differentiated, but the database indicated that the expression was testis-specific (**Figure 6**). The NCBI database also indicated the testis-specific expression of FAM205A (not shown). The expression pattern of FAM205A was also found in the Human Protein Atlas, in which FAM205A was expressed in testis and in no other tissues examined at the mRNA level (not shown), confirming our PCR-based data. According to the Human Protein Atlas, cells in the seminiferous ducts (sperm and immature sperm cells) of the testis were clearly detected, but Leydig cells were not stained immunohistochemically (**Figure** 7). As mentioned in the Human Protein Atlas, staining was clearly detected in acrosomes in spermatids (**Figure** 7). Considering that the antibody used in the Human Protein Atlas could not differentiate FAM205A and FAM75 (because a recombinant C-terminal 104 aa fragment that is almost identical in both FAM205A and FAM75 was used as an antigen), both proteins were likely stained in the tissue sections.

3.8 Structural and functional predictions

We performed several sequence analyses to characterize the sequences of FAM75 (**Figure 8**). When FAM75 and FAM205A were subjected to SOSUI, the

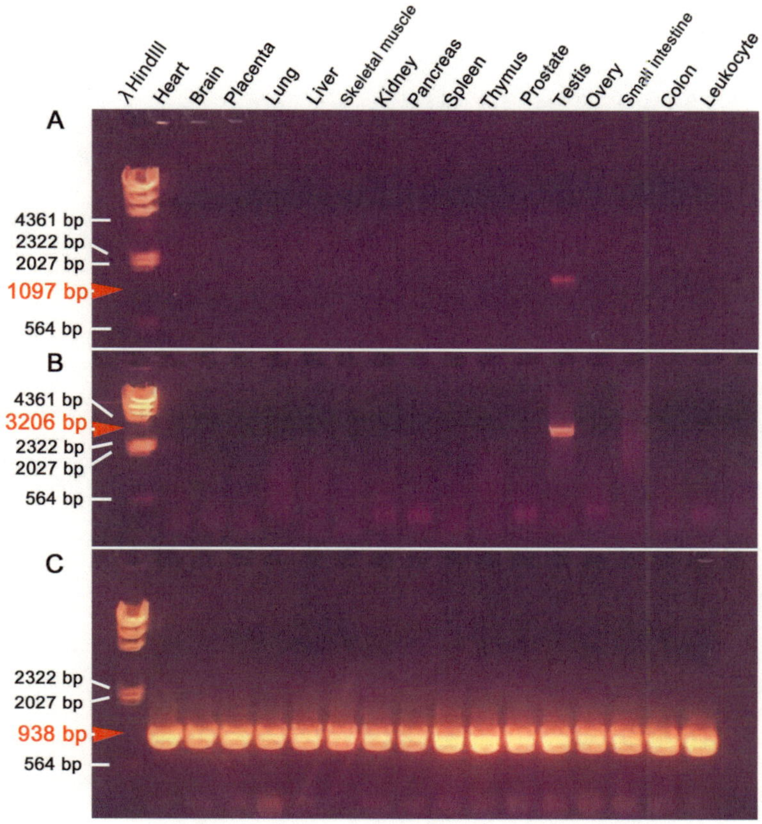

Figure 5.
PCR products from human-tissue cDNA templates. (A) FAM75 and FAM205A. Primers were designed to amplify both FAM75 and FAM205A. A DNA fragment with the expected size (1097 bp) was amplified only from testis. (B) FAM205A. Primers were designed to amplify only FAM205A. A DNA fragment with the expected size (3206 bp) was amplified only from testis. (C) G6PDH as a positive control. A DNA fragment with the expected size (938 bp) was amplified from all tissues tested.

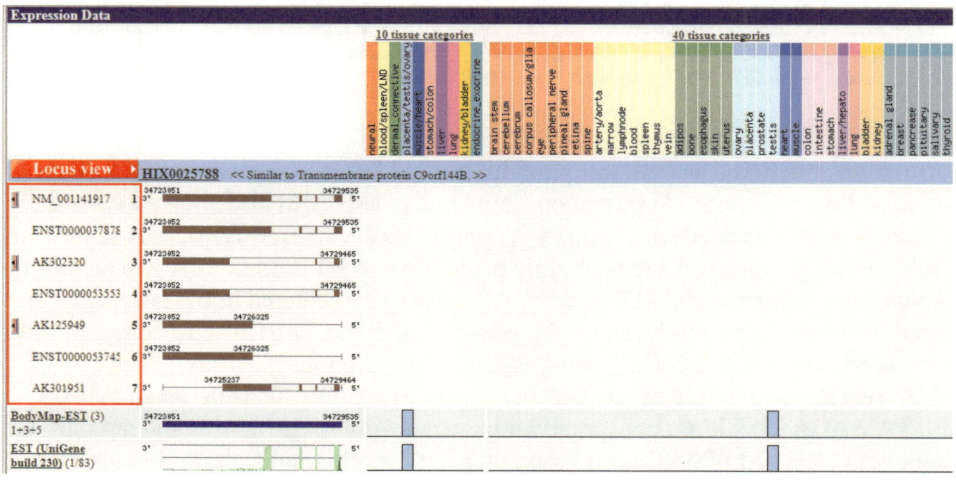

Figure 6.
Gene expression profile of FAM205A in human tissues (H-ANGEL). NM_001141917 indicates FAM205A, and AK125949 indicates FAM75. http://www.h-invitational.jp/hinv/h-angel/wge_server.cgi?gpid=HIX0025788.

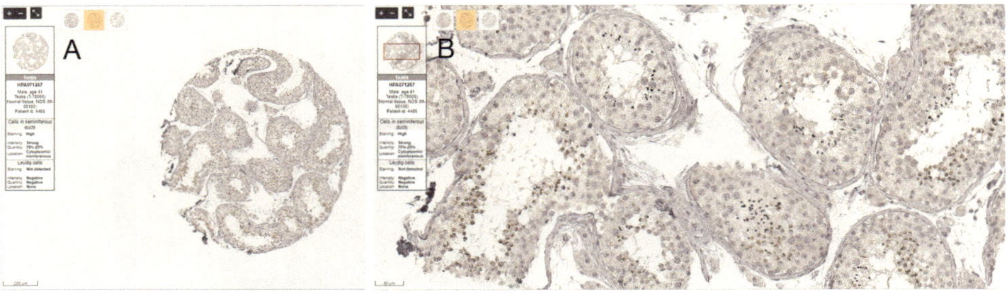

Figure 7.
Immunohistochemical detection of FAM205A in a normal human testis section. FAM75 is also likely stained if it is present, because the antibody was raised against a recombinant C-terminal 104-aa fragment, which is found both in FAM205A and FAM75. Pictures were taken from the Human Protein Atlas (HPA071267, male, age 41, testis (T-78000), normal tissue, NOS (M-00100), patient ID: 4485). Dark brown signals are seen in spermatids and other differentiating cells in the seminiferous ducts. The crescent-like staining likely represents developing acrosomes. Two additional samples (ages 25 and 65) are shown in the Human Protein Atlas with essentially the same results. (A) Entire section. (B) High magnification of A.

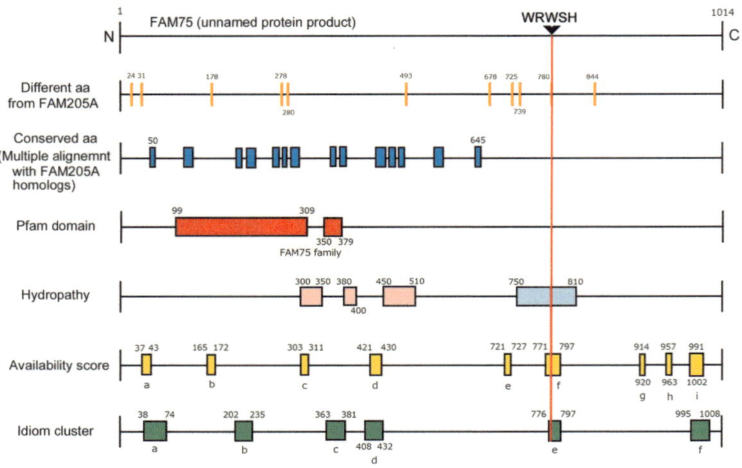

Figure 8.
Identification of possible functional sites in FAM75. Shown are different amino acids from FAM205A (NCBI GenBank record), conserved amino acids based on multiple alignment with FAM205A homologs, Pfam domain, hydropathy, availability score, and idiom clusters. The location of the amino acid H in the candidate human-specific pentat WRWSH is indicated by a vertical red line.

former was predicted as a soluble protein, but the latter was predicted as a membrane protein with a single transmembrane helix. TMHMM also showed essentially the same results. Indeed, the Human Protein Atlas considered FAM205A to be both a membrane protein and a cytoplasmic protein based on immunohistochemical results, suggesting that FAM75 and FAM205A may be detected in the cytoplasm and in membranes, respectively, as predicted by SOSUI and TMHMM. In contrast, both were predicted to be "nuclear" using PSORT II Prediction.

To search for possible functional sites, different amino acids between FAM75 and FAM205A (**Table 1**), conserved amino acids among FAM205A and similar sequences (top 100 BLAST data) based on multiple alignment, Pfam domain data, a hydropathy plot, an availability plot, and an idiom plot were aligned together (**Figure 8**). Conserved amino acids were located mostly in the N-terminal side, in which the "FAM75 domain" identified by Pfam was also located. WRWSH was located at the center of the hydrophilic region in the C-terminal side and corresponded to a high availability region and a high idiom region, although their significance was not clear at this point.

4. Discussion

In this paper, we identified a WRWSH-containing protein, FAM75, as a candidate human-specific protein. We assumed that pentats with high availability scores in humans and no occurrence ($A = -1$) in nine other mammals might be contained in a human-specific protein. The current method based on this assumption indeed identified FAM75. Although the DNA sequence coding for WRWSH is one of the SNP variants in the human genome (i.e., WRWSH was not conserved in all human populations), this fact does not exclude the candidacy of WRWSH as a human-specific pentat, because we do not know when this SNP variant was created during human history. Likely, not all point mutations are functionally equal; some point mutations may incidentally create a rare pentat like WRWSH that may contribute to functional novelty. Interestingly, the FAM75 transcript was found only in humans as an alternative splicing transcript of FAM205A. In this sense, our SCS-based search for human-specific proteins successfully identified what we wanted to identify. The success of this study may simply be fortunate. On the other hand, there are many other candidate human-specific pentats that we did not examine in detail. Changing search conditions, including the length of amino acid sequences (i.e., triplets, quartets, and longer SCSs), could identify further candidate human-specific SCSs.

The present study showed that the SCS-based approach is a relevant addition to a list of practical sequence comparison methods. As with other methods, the SCS-based method is influenced by SNPs, accuracy, and the amount of information in databases. For example, the human genome has numerous SNP variations, and there is much less genomic information for other primates than for humans. A and ΔA scores, which were used to search in this study, are dependent on databases. WRWSH had high ΔA between humans and nine other mammals, and this is partly because there were many human protein records that contained this pentat at the time of the database search. Unfortunately, most of these records were later removed from human databases (NCBI GenBank records) because of the uncertainty of their status (although they were not rejected completely). This illustrates the importance of database quality in genome comparison studies. However, whatever ΔA was, we focused on the pentats that were not used at all ($A = -1$) in the nine other nonhuman mammals, which made the choice of pentats for further investigation less sensitive to database quality.

FAM75 and FAM205A appear to be alternative splicing products from the same genomic locus in humans (**Figure 1**). The relationship of the evolutionary invention of FAM75 as an alternative splicing product with that of a SNP variant for WRWSH is unclear. We cannot exclude the possibility that this may be a simple coincidence, but this coincidence is in accordance with our starting hypothesis for this study: proteins containing a human-specific pentat may indeed be human-specific as proteins. We confirmed the expression of FAM205A and/or FAM75 at the mRNA level in human tissues (**Figure 5**). At the protein level, the FAM205A protein (and probably also the FAM75 protein) was shown to be located in cells in seminiferous ducts and in acrosomes in spermatids in the testis (**Figure 7**). Interestingly, FAM205A was also detected in the human sperm nucleus in a proteomic study [75]. Although it is difficult to distinguish FAM75 and FAM205A at the mRNA and protein levels, it is demonstrated that the FAM75/FAM205A gene is not a pseudogene, and protein products are actively produced in testis. The discovery of the IRES element in FAM75 mRNA also supports the idea that FAM75 mRNA is actively translated into proteins. On the other hand, we found two additional alternative splicing products in H-InvDB (**Figure 1**). These additional mRNAs were not examined in this paper, because of insufficient information. However, their status is of interest if they really exist; they may have similar but slightly different functions from FAM205A and FAM75.

Mechanistically, alternative splicing may be a relatively easy way to create a new protein sequence. It may be considered not only a "regulatory change" (according to the regulatory hypothesis, because the evolutionary invention of a new alternative splicing product conserves the original protein-coding DNA sequence and gene function and thus is more conservative with respect to species evolution) but also a "sequence change" (according to the constituent hypothesis, because the protein sequence is changed). These two modes are likely intermingled in this case. To extrapolate this argument, transcriptome studies of alternative splicing or RNA processing may be fruitful to identify human-specific genes. The present discovery of the IRES element in the FAM75 mRNA may be surprising because IRES elements are mostly viral, and cellular elements are relatively rare [72–74]. A search for IRES elements in the genome may also be fruitful.

The evolution of WRWSH and FAM75 in relation to human speciation is an important but uncertain aspect to be discussed. There are two kinds of "human-specific" proteins. First, a group of proteins may have been involved in the early step of speciation of *Homo sapiens* from its ancestral species. Second, after the establishment of *Homo sapiens*, additional changes in a group of proteins may occur as a reinforcement process. In either case, these proteins may be called human-specific. If the pentat WRWSH (or FAM75) played a role in these early or late steps of human speciation, this pentat is human-specific, and it would be later mutated back to WRWSR in African populations. In this case, WRWSH was once assimilated completely in the human population during speciation, and a new WRWSR sequence is now assimilating, as genetic assimilation has been considered a key process in species evolution [76–78]. However, because WRWSH is relatively rare in African populations, it is more parsimonious to think that WRWSH evolved after human speciation in Asian or European populations. We speculate that FAM75 may have been invented from FAM205A to play a role in human speciation, but at least in the early stage, FAM75 exclusively contained WRWSR, as in the other great apes. WRWSH may then have been invented in FAM75 to reinforce human speciation. Alternatively, WRWSH did not play any role in human speciation, and its reinforcement simply fortified the function of FAM75 in some populations relatively recently.

What is the function of FAM75 in human testes? According to the results of immunohistochemistry (the Human Protein Atlas), SOSUI, and TMHMM, we speculate that FAM75 appears to function differently from FAM205A in different cellular sites. Because FAM75 is likely located in acrosomes (**Figure 7**), this protein may be involved in the process of fertilization. A possibility is that FAM75 confers human specificity to prevent cross-species fertilization with ancestral species. The FAM75/FAM205A genomic locus in humans has an additional two alternative splicing products, which were not pursued in the present study, and orangutans and mice appeared to have three transcripts from the same locus. It is tempting to speculate that this locus partly contributes to speciation in primates and other mammals by restricting cross-species fertilization in ancestral species.

Molecularly, the main function of FAM75 may be located in the "FAM75 domain" located at the N-terminal side of the molecule (**Figure 8**), but because WRWSH is located in a hydrophilic region at the C-terminal side of the molecule, this hydrophilic region may function in human specificity. Indeed, the conserved regions are mostly located at the N-terminal side, probably for the general function of FAM75. The hydrophilic region also coincides with high availability and idiom-cluster regions.

Testis is known to be the tissue of the fastest evolution among other tissues based on gene expression comparisons in mammals, including the great apes [13, 14, 16–18, 79]. This flexibility may reflect diverse species-specific sexual behaviors. Mating is nonselective and frequent in chimpanzees, and only the highest-ranked male can mate in gorillas [80]. These behaviors have been thought to be related to testis-size differences; the chimpanzee has relatively large testes, and the gorilla has small ones [80]. Human testis size lies between these extremes, which may be related to the molecular evolution of FAM75 to modulate sperm development in testes or to withstand moderate sperm competition.

A recent finding that the gene locus for FAM205A is a susceptible locus for intracerebral hemorrhage (ICH) [81] is somewhat surprising. Either FAM205A or FAM75 may be expressed in cerebral cells at low levels or in restricted regions of the brain. It is tempting to speculate that a pleiotropic protein for both fertilization and brain development, such as FAM75/FAM205A, might have played a role in human evolution. The fact that the FAM205A/FAM75 gene is located not in a sex chromosome but in chromosome 9, despite its expression in the testis, might further suggest its dual role in sexual and nonsexual aspects of human specificity.

5. Conclusions

Our SCS-based approach identified FAM75, a WRWSH-containing protein, as a candidate human-specific protein. Its uniqueness in humans may be acquired not only by a point mutation for WRWSH but also by novel alternative splicing. Together with FAM205A, FAM75 is likely expressed in human testis, and its possible expression in acrosomes suggests its potential function in fertilization and thus in human speciation. Its potential pleiotropic function in the brain is very interesting and may also be investigated in the future.

Acknowledgements

We thank Miki Kawauchi, Motosuke Tsutsumi, Hideka Konno, and other members of the BCPH Unit of Molecular Physiology for technical assistance and discussions. This work was supported by the Sekisui Chemical Grant Program for Research to JMO. This work was also supported by basic funds to JMO and MN from the University of the Ryukyus.

Conflict of interest

Authors declare no competing interests.

Author details

Shiho Endo[1], Kenta Motomura[2], Masakazu Tsuhako[1], Yuki Kakazu[2], Morikazu Nakamura[2] and Joji M. Otaki[1*]

1 The BCPH Unit of Molecular Physiology, Department of Chemistry, Biology and Marine Science, Faculty of Science, University of the Ryukyus, Okinawa, Japan

2 Department of Information Science, Faculty of Engineering, University of the Ryukyus, Okinawa, Japan

*Address all correspondence to: otaki@sci.u-ryukyu.ac.jp

References

[1] Leigh SR. Brain growth, life history, and cognition in primate and human evolution. American Journal of Primatology. 2004;**62**:139-162

[2] Gilbert SF, Epel D. Ecological Developmental Biology. 2nd ed. Sunderland, MA: Sinauer Associates; 2015

[3] King M-C, Wilson AC. Evolution at two levels in humans and chimpanzees. Science. 1975;**188**:107-116

[4] Lander ES et al. Initial sequencing and analysis of the human genome. Nature. 2001;**409**:860-921

[5] Mikkelsen T et al. Initial sequence of the chimpanzee genome and comparison with the human genome. Nature. 2005;**437**:69-87

[6] Locke DP et al. Comparative and demographic analysis of orangutan genomes. Nature. 2011;**469**:529-533

[7] Scally A et al. Insights into hominid evolution from the gorilla genome sequence. Nature. 2012;**483**:169-175

[8] Vark A, Geschwing DH, Eichler EE. Explaining human uniqueness: Genome interactions with environment, behaviour, and culture. Nature Review Genetics. 2008;**9**:749-763

[9] McLean CY et al. Human-specific loss of regulatory DNA and the evolution of human-specific traits. Nature. 2011;**471**:216-219

[10] Enard W et al. Intra- and interspecific variation in primate gene expression patterns. Science. 2002;**296**: 340-343

[11] Preuss TM, Caceres M, Oldham MC, Geschwind DH. Human brain evolution: Insights from microarrays. Nature Review Genetics. 2004;**5**:850-860

[12] Boyd JL, et al. Human-chimpanzee differences in a FZD8 enhancer alter cell-cycle dynamics in the developing neocortex. Current Biology. 2015;**25**: 772-779

[13] Khaitovich P, et al. Parallel patterns of evolution in the genomes and transcriptomes of humans and chimpanzees. Science. 2005;**309**: 1850-1854

[14] Khaitovich P, Enard W, Lachmann M, Pääbo S. Evolution of primate gene expression. Nature Review Genetics. 2006;**7**:693-702

[15] Nshon J-L. Birth of 'human-specific' genes during primate evolution. Genetica. 2003;**118**:193-208

[16] Kaessmann H. Origins, evolution, and phenotypic impact of new genes. Genome Research. 2010;**20**:1313-1326

[17] Nielsen R, et al. A scan for positively selected genes in the genomes of humans and chimpanzees. PLoS Biology. 2005;**3**:e170

[18] Tay SK, Blythe J, Lipovich L. Global discovery of primate-specific genes in the human genome. Proceedings of the National Academy of Sciences of the United States of America. 2009;**106**: 12019-12024

[19] Kronenberg ZN et al. High-resolution comparative analysis of great ape genomics. Science. 2018;**360**:eaar6343

[20] Enard W, et al. Molecular evolution of FOXP2, a gene involved in speech and language. Nature. 2002;**418**:869-872

[21] Atkinson EG, et al. No evidence for recent selection at *FOXP2* among diverse human populations. Cell. 2018; **174**:1424-1435.e15

[22] Altschul SF, Gish W, Miller W, Myers EW, Lipman DJ. Basic local alignment search tool. Journal of Molecular Biology. 1990;**215**:403-410

[23] Kapustin Y, Souvorov A, Tatusova T, Lipman D. Splign: Algorithms for computing spliced alignments with identification of paralogs. Biology Direct. 2008;**3**:20

[24] Vinga S, Almeida JS. Alignment-free sequence comparison—A review. Bioinformatics. 2003;**19**:513-523

[25] Zielezinski A, Vinga S, Almeida J, Karlowski WM. Alignment-free sequence comparison: Benefits, applications, and tools. Genome Biology. 2017;**18**:186

[26] Otaki JM, Firestein S. Length analyses of mammalian G-protein-coupled receptors. Journal of Theoretical Biology. 2001;**211**:77-100

[27] Otaki JM, Mori A, Itoh Y, Nakayama T, Yamamoto H. Alignment-free classification of G-protein-coupled receptors using self-organizing maps. Journal of Chemical Information and Modeling. 2006;**46**:1479-1490

[28] Otaki JM, Ienaka S, Gotoh T, Yamamoto H. Availability of short amino acid sequences in proteins. Protein Science. 2005;**14**:617-625

[29] Otaki JM, Gotoh T, Yamamoto H. Potential implications of availability of short amino acid sequences in proteins: An old and new approach to protein decoding and design. Biotechnology Annual Review. 2008;**14**: 109-141

[30] Otaki JM, Tsutsumi M, Gotoh T, Yamamoto H. Secondary structure characterization based on amino acid composition and availability in proteins. Journal of Chemical Information and Modeling. 2010;**50**:690-700

[31] Tsutsumi M, Otaki JM. Parallel and antiparallel β-strands differ in amino acid composition and availability of short constituent sequences. Journal of Chemical Information and Modeling. 2011;**51**:1457-1464

[32] Motomura K, Fujita T, Tsutsumi M, Kikuzato S, Nakamura M, Otaki JM. Word decoding of protein amino acid sequences with availability analysis: A linguistic approach. PLoS One. 2012;**7**: e50039

[33] Motomura K, Nakamura M, Otaki JM. A frequency-based linguistic approach to protein decoding and design: Simple concepts, diverse applications, and the SCS package. Computational and Structural Biotechnology Journal. 2013;**5**: e201302010

[34] Bresell A, Persson B. Characterization of oligopeptide patterns in large protein sets. BMC Genomics. 2007;**8**:346

[35] Tuller T, Chor B, Nelson N. Forbidden penta-peptides. Protein Science. 2007;**16**:2251-2259

[36] Figureau A, Soto MA, Tohá J. A pentapeptide-based method for protein secondary structure prediction. Protein Engineering. 2003;**16**:103-107

[37] Pe'er I, Felder CE, Man O, Silman I, Sussman JL, Beckmann JS. Proteomic signatures: Amino acid and oligopeptide compositions differentiate among phyla. Proteins. 2004;**54**:20-40

[38] Poznański J, et al. Global pentapeptide statistics are far away from expected distributions. Scientific Reports. 2018;**8**:15178

[39] Patel A, et al. Pentamers not found in the universal proteome can enhance antigen specific immune responses and adjuvant vaccines. PLoS One. 2012;**7**: e43802

[40] Navon SP, et al. Amino acid sequence repertoire of the bacterial proteome and the occurrence of untranslatable sequences. Proceedings of the National Academy of Sciences of the United States of America. 2016;**113**: 7166-7170

[41] Zemková M, Zahradnik D, Mokrejš M, Flegr J. Parasitism as the main factor shaping peptide vocabularies in current organisms. Parasitology. 2017;**144**:975-983

[42] Burdukiewicz M, Sobczyk P, Rödiger S, Duda-Madej A, Mackiewicz P, Kotulska M. Amyloidogenic motifs revealed by n-gram analysis. Scientific Reports. 2017;**7**:12961

[43] Vries JK, Liu X, Bahar I. The relationship between n-gram patterns and protein secondary structure. Proteins. 2007;**68**:830-838

[44] Daeyaert F, Moereels H, Lewi PJ. Classification and identification of proteins by means of common and specific amino acid n-tuples in unaligned sequences. Computer Methods and Programs in Biomedicine. 1998;**56**:221-233

[45] Chou KC. Prediction of protein cellular attributes using pseudo-amino acid composition. Proteins. 2001;**43**: 246-255

[46] Chou KC, Cai YD. Predicting protein quaternary structure by pseudo amino acid composition. Proteins. 2003; **53**:282-289

[47] Cai YD, Chou KC. Predicting enzyme subclass by functional domain composition and pseudo amino acid composition. Journal of Proteome Research. 2005;**4**:967-971

[48] Popov O, Segal DM, Trifonov EN. Linguistic complexity of protein sequences as compared to texts of human language. BioSystems. 1996;**38**:65-74

[49] Eroglue S. Language-like behavior of protein length distribution in proteomes. Complexity. 2014;**20**:12-21

[50] de Brevern AG, Valadié H, Hazout S, Etchebest C. Extension of a local backbone description using a structural alphabet: A new approach to the sequence-structure relationship. Protein Science. 2002;**11**:2871-2886

[51] de Brevern AG. New assessment of a structural alphabet. In Silico Biology. 2005;**5**:283-289

[52] Joseph AP, et al. A short survey on protein blocks. Biophysical Reviews. 2010;**2**:137-145

[53] de Brevern AG, Joseph AP. Species specific amino acid sequence-protein local structure relationships: An analysis in the light of a structural alphabet. Journal of Theoretical Biology. 2011;**276**: 209-217

[54] Nekrasov AN, et al. A minimum set of stable blocks for rational design of polypeptide chains. Biochimie. 2019; **160**:88-92

[55] Kakazu Y, Nakamura M, Otaki JM. Idiom networks for short constituent sequences of amino acids. In: 2015 International Conference on Intelligent Informatics and Biomedical Sciences (ICIIBMS). 2005. pp. 15-19

[56] Kakazu Y, Nakamura M, Otaki JM. GPU acceleration for availability scoring of short constituent amino acid sequences. In: 2015 Third International Symposium on Computing and Networking (CANDAR). 2015. pp. 598-600

[57] Takeda J, et al. H-InvDB in 2013: An omics study platform for human functional gene and transcript discovery. Nucleic Acids Research. 2013;**41**:D915-D919

[58] Kumar S, Stecher G, Tamura K. MEGA7: Molecular evolutionary genetics analysis version 7.0 for bigger datasets. Molecular Biology and Evolution. 2016;**33**:1870-1874

[59] Huang HY, Chien CH, Jen KH, Huang HD. RegRNA: A regulatory RNA motifs and elements finder. Nucleic Acids Research. 2006;**34**: W429-W423

[60] Kitts A, Sherry S. The single nucleotide polymorphism database (dbSNP) of nucleotide sequence variation. In: McEntyre J, Ostell J, editors. The NCBI Handbook. Bethesda: National Center for Biotechnology Information. Chapter 5; 2002

[61] Sudmant PH et al. An integrated map of structural variation in 2504 human genomes. Nature. 2015;**526**:75-81

[62] Fagerberg L et al. Analysis of the human tissue-specific expression by genome-wide integration of transcriptomics and antibody-based proteomics. Molecular & Cellular Proteomics. 2014;**13**:397-406

[63] Uhlén M et al. Tissue-based map of the human proteome. Science. 2015;**347**: 1260419

[64] Hirokawa T, Boon-Chieng S, Mitaku S. SOSUI: Classification and secondary structure prediction system for membrane proteins. Bioinformatics. 1998;**14**:378-379

[65] Krogh A, Larsson B, von Heijne G, Sonnhammer ELL. Predicting transmembrane protein topology with a hidden markov model; application to complete genomes. Journal of Molecular Biology. 2001;**305**:567-580

[66] Nakai K, Horton P. PSORT: A program for detecting sorting signals in proteins and predicting their subcellular localization. Trends in Biochemical Sciences. 1999;**24**:34-35

[67] El-Gebali S, et al. The Pfam protein families database in 2019. Nucleic Acids Research. 2018;**47**:D427-D432

[68] Rice P, Longden I, Bleasby A. EMBOSS: The European molecular biology open software suite. Trends in Genetics. 2000;**16**:276-277

[69] NCBI. *Homo sapiens* Annotation Release 109. Date of submission of annotation to the public databases: March 26, 2018. Available at: www.ncbi.nlm.nih.gov/genome/annotation_euk/Homo_sapiens/109/

[70] NCBI. *Homo sapiens* Annotation Release 106. Date of submission of annotation to the public databases: February 3, 2014. Available at: www.ncbi.nlm.nih.gov/genome/annotation_euk/Homo_sapiens/106/

[71] Humphray SJ et al. DNA sequence and analysis of human chromosome 9. Nature. 2004;**429**:369-374

[72] Lozano G, Francisco-Velilla R, Martinez-Salas E. Deconstructing internal ribosome entry site elements: An update of structural motifs and functional divergences. Open Biology. 2018;**8**:180155

[73] Du X, et al. Second cistron in *CACNA1A* gene encodes a transcription factor mediating cerebellar development and SCA6. Cell. 2013;**154**: 118-133

[74] Xue S, Tian S, Fujii K, Kladwang W, Das R, Barma M. RNA regulons in *Hox* 5' UTRs confer ribosome specificity to gene regulation. Nature. 2015;**517**:33-38

[75] de Mateo S, Castillo J, Estanyol JM, Ballescà JL, Oliva R. Proteomic characterization of the human sperm nucleus. Proteomics. 2011;**11**:2714-2726

[76] Waddington CH. Genetic assimilation of the bithorax phenotype. Evolution. 1956;**10**:1-13

[77] Otaki JM, Hiyama A, Iwata M, Kudo T. Phenotypic plasticity in the range-margin population of the lycaenid butterfly *Zizeeria maha*. BMC Evolutionary Biology. 2010;**10**:252

[78] Hiyama A, Taira W, Otaki JM. Color-pattern evolution in response to environmental stress in butterflies. Frontiers in Genetics. 2012;**3**:15

[79] Brawand D, et al. The evolution of gene expression levels in mammalian organs. Nature. 2011;**478**:343-348

[80] Harcourt AH, Harvey PH, Larson SG, Short RV. Testis weight, body weight and breeding system in primates. Nature. 1981;**293**:55-57

[81] Yamada Y, et al. Identification of nine genes as novel susceptibility loci for early-onset ischemic stroke, intracerebral hemorrhage, or subarachnoid hemorrhage. Biomedical Reports. 2018;**9**:8-20

Section 3

Bioinformatics and the Related Software Tools and Databases

Chapter 3

Bioinformatics as a Tool for the Structural and Evolutionary Analysis of Proteins

Edna María Hernández-Domínguez, Laura Sofía Castillo-Ortega, Yarely García-Esquivel, Virginia Mandujano-González, Gerardo Díaz-Godínez and Jorge Álvarez-Cervantes

Abstract

This chapter deals with the topic of bioinformatics, computational, mathematics, and statistics tools applied to biology, essential for the analysis and characterization of biological molecules, in particular proteins, which play an important role in all cellular and evolutionary processes of the organisms. In recent decades, with the next generation sequencing technologies and bioinformatics, it has facilitated the collection and analysis of a large amount of genomic, transcriptomic, proteomic, and metabolomic data from different organisms that have allowed predictions on the regulation of expression, transcription, translation, structure, and mechanisms of action of proteins as well as homology, mutations, and evolutionary processes that generate structural and functional changes over time. Although the information in the databases is greater every day, all bioinformatics tools continue to be constantly modified to improve performance that leads to more accurate predictions regarding protein functionality, which is why bioinformatics research remains a great challenge.

Keywords: computational biology, databases, proteomics, transcriptomics, functional genomics, phylogeny

1. Introduction

The study to understand the functioning of the cell, as well as the molecules and processes that are carried out within it, originated the use of various disciplines and sciences to facilitate the progress in research for its characterization over time. In the 1950s, the sequencing of small biological molecules began, and in 1956, the sequencing of the first protein was achieved. Thus, Margaret O. Dyhoff determined that bovine insulin is a small peptide of 51 amino acids. With these advances and the constant production of biological information, there was a need to collect and organize all the information generated from these sequencing projects [1]. In 1965, the first biological sequence database was created, in which all the DNA and protein sequences described up to that time were stored and made available to the scientific community. Eight years later, the oldest known database was created, which is still in force today, *Protein Data Bank* (PDB) [2].

In the 80s, bioinformatics had already gained a new meaning in scientific research, so several research groups such as Theoretical Biology and Biophysics Group attached to the American Institute The Alamos National Laboratory, together with Stanford University, gave rise to the best-known database in the world called GenBank. Almost at the same time, in 1981, Temple Smith and Michael Waterman extensively reviewed the mathematical algorithms for comparing biological sequences. As a result of their analysis, they generated the well-known local alignment algorithm that allowed to optimize the comparison of biological sequences, being the most important contribution for the direct comparison of sequences and cornerstone of the alignment by sequence pair [3].

A few years after the creation of *GenBank*, its European and Asian versions were generated, known as the EMBL database (*European Molecular Biology Laboratory*) and DDBJ (*DNA Data Bank of Japan*) in 1981 and 1984, respectively. In 1985 the FASTA algorithm (*FAST-AII*) of sequence comparisons was reported, which operated as a search engine for similar sequences within the *GenBank* [4]. During the years from 1987 to 1990, databases for protein sequences were propelled which resulted in the creation of Swiss-Prot and PIR (Protein Information Resource). In 1990, another of the most important milestones in bioinformatics originated the BLAST algorithm (Basic Local Alignment Tool) that completely revolutionized the exploration and search of biological sequences in databases [5].

The National Center for Biotechnology Information (NCBI) makes the following definition:

Bioinformatics is a field of science in which various disciplines such as applied mathematics, statistics, artificial intelligence, chemistry, biochemistry, computing and information technology converge, whose objective is to facilitate the discovery of new biological ideas, as well as create global perspectives from which unifying principles in biology can be discerned [6].

It consists of two complementary subfields with each other:

1. The development of computer tools and databases.

2. The application of these in the generation of biological knowledge to better understand living systems [7].

According to the *National Institute of Health* of the United States, bioinformatics or also called computational biology, deals with the development and application of analytical data and theoretical methods, mathematical modeling and computer simulation techniques to study biological, behavioral and social systems [8]. The programs use public or private databases (with restricted access or with economic value) that have been created with information that is constantly growing and managed by institutions from various sectors. The main databases used in computational biology are described below:

1.1 Biological databases

- *Primary databases* contain original biological data. They are raw sequence files or structural data (for example, *GenBankm y Protein Data Bank)* [6].

- *Secondary databases* contain information processed computationally based on primary data. Translated protein sequence databases contain the functional annotation belonging to this category (for example, *Swiss-Prot and PIR)* [6].

- *Specialized databases* are those that serve a particular research interest (for example, *Flybase*). The HIV sequence database and Ribosomal Database Project are examples of databases that specialize in a particular organism or a certain type of data. Many of the problems detected in scientific research lie in the need to connect secondary and specialized databases to primary databases. It is desirable that entries in a database be cross-referenced or linked to related entries in other databases that contain additional information [6].

There are primary databases, which contain direct information on the sequence, structure or pattern of DNA or protein expression, and secondary, which contains data derived from primary databases, such as mutations, evolutionary relationships, grouping by families or by functions, involvement in diseases, etc.

1.2 Databases for protein analysis (amino acid sequence databases)

Swiss-Prot: It contains annotated or commented sequences, that is, each sequence has been reviewed, documented and linked to other databases. External link: Swiss-Prot in the EBI (*http://www.ebi.ac.uk/swissprot/access.html*), Swiss-Prot in ExPASy (*http://us.expasy.org/sprot/*) [9].

TrEMBL: *Translation of EMBL Nucleotide Sequence Database* includes the translation of all coding sequences derived from (EMBL-BANK) and which have not yet been annotated in Swiss-Prot. External link: TrEMBL (*http://www.ebi.ac.uk/trembl/*) [9].

PIR: *Protein Information Resource* is divided into four sub-bases that have a decreasing annotation level. External link: PIR (*http://pir.georgetown.edu/*) [9].

ENZYME: It links the complete enzyme activity classification to the Swiss-Prot sequences. External link: ENZYME (*http://us.expasy.org/enzyme/*) [9].

PROSITE: It contains information on the secondary structure of proteins, families, domains, etc. External link: PROSITE (*http://us.expasy.org/prosite/*) [9].

INTERPRO: It integrates information from various secondary structure databases such as PROSITE, providing links to other databases and more extensive information. External link: INTERPRO (*http://www.ebi.ac.uk/interpro/index.html*) [9].

PDB: *Protein Data* Bank is the 3-D tertiary structure database of proteins that have been crystallized. External link: PDB (*http://www.rcsb.org/pdb/*) [9].

1.3 Data warehouse

A *Data Warehouse* (*DW*) is a set of integrated data oriented to a subject, which vary over time and are not transitory, which support the decision-making process of the administration [10]. From the review of the bioinformatics projects it is found that the requirements of this field require the storage of large volumes of data, with multiple dimensions, of extended periods of time and with heterogeneous formats as well as their sources. For example, *Ligand Depot* is an integrated data source for finding information about small molecules, proteins and nucleic acids. It focuses on providing chemical and structural information for small molecules. Accepts keyword-based queries, also provides a graphical interface for conducting chemical substructure searches, and allows access to a wide variety of web resources [11].

1.4 Data mining in bioinformatics

Data mining is oriented towards the study of techniques to extract valuable information from a large amount of biological data. For this, efficient software tools are necessary to recover data, compare biological sequences, discover patterns and visualize the discovery of knowledge [8].

Among the most common data mining techniques in bioinformatics can be highlighted [8]:

KDD is the complete process of extracting knowledge, not trivial, previously unknown and potentially useful from a data set.

KDT is oriented to the extraction of knowledge from data (unstructured in natural language) stored in textual databases, is identified with the discovery of knowledge in the texts.

1.5 Applications of bioinformatics

The areas in which bioinformatics is currently developed are many and varied, ranging from simple tasks such as direct acquisition of data from DNA or protein sequencing assays (when techniques such as mass spectrophotometry are used), until the development of software for the storage and analysis of the data, which implies in many cases, the generation of algorithms that require both mathematical and biological knowledge. Within the areas in which bioinformatics takes place are genomics, proteomics, pharmacogenetics and phylogeny. The plant genome databases and gene expression analysis of this profile have played an important role in the development of new crop varieties that have higher productivity and more disease resistance [7].

Specifically, bioinformatics encompasses the development of databases or knowledge to store and retrieve biological data, algorithms to analyze and determine their relationships with biological data, and the statistical tools to identify and interpret data sets. The following describes in detail what refers to metabolomics, transcriptomics, proteomics, comparative genomics, functional genomics, phylogeny and protein modeling.

2. Metabolomic data analysis

The metabolomics was originally proposed as a tool of functional genomics, but its use has been extended much more, as it has had great advances like other omics sciences, such as transcriptomics and proteomics; because the metabolomic work is determined by physical-chemical characteristics of organic molecules unlike the genes, mRNA and proteins that come from a specific sequence, so the success of the characterization of these biopolymers is thanks to bioinformatics technology and tools that help sequence characterization [12]. Its objective is to detect, quantify and interpret the overall analysis of all metabolites; these studies are used in various areas and, like proteomics, one of its main contributions is biomarkers, helping to identify metabolites that are correlated with diseases and environmental exposures [13]. Metabolites are chemical entities that do not come from a transfer of information within the cell, coupled with this, they are also characterized by being diverse as they are substrates and metabolism products that drive essential cellular functions, such as energy production and storage, signal transduction and cell apoptosis; in this great diversity of chemical structures we find endogenous and exogenous metabolites, the former are produced naturally by an organism and the latter come from interaction with the outside. The great diversity of molecules reflects in a wide range of polarities, molecular weights, functional groups, stability and chemical reactivity, etc. [12, 13].

Among the first reports of metabolite detection are those where mass spectrometry (MS) was used to separate a wide range of metabolites present in urine and tissue extracts [14]. In addition, multicomponent analyzes were described to obtain the metabolic profile for three types of urinary constituents: steroids, acids, drugs and drug metabolism [15]. On the other hand, there are reports where physical, chemical or psychological changes can cause biological responses such as oxidative stress and inflammation; among the biomarkers that are the result of a chemical reaction are lipoperoxides or oxidized proteins that are the result of the reaction of molecules with reactive oxygen species (ROS) and those that represent the biological response to stress, such as the transcription factor NRF2 or inflammation and inflammatory cytokines [16]. Among the best known and clinically used examples we find glucose as a marker of diabetes [17] and phenylalanine as a marker of congenital metabolic disorder [18].

Because metabolites play important roles in the biological pathways; its differential flow or regulation can reveal new knowledge about diseases and environmental influences, so one of the most important objectives of the metabolic analysis has been to assign the identity of the metabolite within a metabolic pathway [19, 20]; generating a large amount of data; requiring for its processing an arduous mathematical, statistical and bioinformatic work [12, 21, 22], this last area is crucial for the development of metabolomics as it helps in the handling of data and information, analytical data processing, metabolomic standards, ontology, statistical analysis, mining and data integration, and mathematical modeling of metabolomic networks with antecedents of biological systems [12], it is also necessary to decide which metabolites are biologically more significant. This can be achieved by helping the identification process, reducing the redundancy of characteristics, presenting better candidates for the MS, accelerating or automating the workflow, recovering data through characteristics through meta-analysis or multigroup analysis, or using stable isotopes and mapping of pathways. For all the above, in recent years, the technologies for analyzing metabolites have undergone improvements, establishing more efficient protocols for experimental design, as well as better sample extraction techniques and data acquisition that have been worthwhile in providing sets of complex and solid data [20].

The database management system for metabolomics requires the collection of raw and processed metadata, some important aspects for comparing data and obtaining results in different laboratories and reproducing experimental conditions are: The nature and treatment of samples prior to study. Among the bases and tools for the analysis and visualization of available data are: Kyoto Encyclopedia of Genes and Genomes (KEGG; http://www.genome.ad.jp/kegg/) [23] and Metabolic Pathways From all Domains of Life (MetaCyc; http://metacyc.org/) [24].

3. Transcriptome data analysis

The genes response to intracellular or extracellular stimuli includes a hierarchy of signals that allows genes encoded in the DNA to be expressed or repressed by the transcription process. The total set of transcripts (RNA molecules) produced by a cell under a given condition and time, is defined as a *transcriptome* [25]. Unlike the genome, the transcriptome is highly dynamic and actively changes as a consequence of factors that influence the stage of development of organisms, as well as the surrounding environmental conditions. In this sense, transcriptomics is an essential tool to interpret the functional elements of the genome, having as object of study, all species of transcripts, messenger RNA, non-coding RNA and small RNAs [26]. Its main purpose being to determine transcriptional structure of genes, that is, where a gene begins and ends (start sites 5′ and 3′ end), posttranscriptional modifications, splicing patterns and differential expression analysis [27].

The RNA molecules synthesized by a cell have a specific function in a given cellular process, the transcripts include: (a) messenger RNA (mRNA) that is the intermediary between the gene information and the proteome. In this way, the amount of mRNA molecules makes it possible to elucidate expression patterns and in turn correlate the abundance of mRNA molecules with changes in protein abundance [28]; (b) non-coding RNA (cRNA) that is responsible for the regulation of gene expression [29]. Determining where, how and when a transcript is generated is essential to know the biological activity of a gene [28]. Analyzing the transcripts that coexist at any given time gives us global information on the cellular state under a certain condition, which has allowed us to establish patterns of gene regulation coordinated with the consequent identification of promoter elements common to several genes [30].

3.1 RNA study technologies and tools in bioinformatic analysis

The RNA study approach has changed from the sequencing of the first determined RNA molecule, to the sequencing of the transcriptome using new generation technologies [25]. *Northern blot* is a technique based on hybridization and radioactive labeling, cDNA microarrays (complementary DNA obtained from mRNA) and cDNA-AFLP tools widely used in studies of expression levels and serial analysis of gene expression (SAGE), at the time they provided relevant information, being Microarrays widely used today [31–35]. However, these techniques require prior knowledge of the genome, have low coverage and are based on hybridization, in this sense the abundance of transcripts is inferred by the intensity of hybridization and the results obtained are noisy, which directly interferes with the reproducibility of the results, besides being insufficient techniques to detect new transcripts [25].

The growing importance of DNA sequencing in model organisms, as well as in the quest to understand the dogma of biology, the NGS technologies (Next Generation Sequencing) arise, which have high yields in the treatment of the sample, are reproducible and highly reliable, as well as accessible and economical, to the point of being more profitable than sequencing by SANGER. These next-generation technologies are based on sequencing by synthesis (SBS) known as pyrosequencing, the transcriptomic variant of pyrosequencing technology is known as short-reading massive parallel sequencing (RNA-seq). The availability of this technology has revolutionized the approach of transcriptome study, having commercially available Roche/454; Applied Biosystems SOLID; HeliScope e Illumina [36].

From the first RNA studies based on sequencing by SANGER to NGS technologies, bioinformatics has been a key tool in the analysis process. Initially the differential expression based on the analysis by Microarrays presented its own computational challenges [36], currently while the reads are shorter than those created by sequencing by SANGER, NGS has a higher performance and generates data set of up to 50 gigabases per run [37], this requires algorithms capable of processing this amount of data in the shortest time possible and with a high degree of reliability.

The study of the transcriptome by RNAseq involves different stages ranging from RNA extraction, library construction, sequencing and data analysis. In this last step four main stages are distinguished (a) *Quality analysis of the reads,* this allows to determine possible problems in the reads. FastQC is a next-generation data quality control tool, which reports graphs and tables providing quality information based on the reads (per base sequence quality); check the quality of subsets of reads (per sequence quality scores); it also shows the proportion of each nucleotide base of the DNA in each base of the reads (per base sequence content); presents the average GC content in the reads and compares that content with the normal distribution (per sequence GC content); shows the proportion of N, that is, unknown nucleotide observed in each reading position (per base N content); shows the size distribution of reads (sequence length distribution); detects adapters in the reads (adapter content); detects possible sequencing problems introduced in the reads after the adapter (k-mer content) https://rtsf.natsci.msu.edu/genomics/tech-notes/fastqc-tutorial-and-faq/ [38]. Is advisable that the length of the reads to be analyzed is the same, also if there is a poor quality in the reads, the procedure to follow is to cut those bases where there is poor quality. Tools such as Fastx-toolkit (https://bio.tools/fastx-toolkit) [39], Trimmomatic [40], PRINSEq [41], Flexbar [42] and others can be used to cut or filter reads, ensuring reliable data for alignment. (b) *Mapping and identification of transcripts*: at this stage the location of the reads with respect to a reference genome is known or a Novo assembly is made. There are three study strategies: (1) the reads are aligned with allocator with gaps to a reference genome (example TopHat, STAR) which allows the identification of new transcripts [43, 44]; (2) If the discovery of new transcripts is not sought, the reads can be aligned to the reference genome using an aligner without gaps for example RSEM [45]; (3) When the genome is not available, the reads are mounted on transcripts what is known as *Novo* assembly (example TRINITY) [46]. In the transcription level analyzes, the isoforms that a gene presents are considered separately. On the contrary, in the level analyzes of gene, the isoforms that it presents form a unit [47]. (c) *Quantification of reads*: Sample reads are quantified in relation to the transcripts that appear in the reference genome or by *Novo* assembly. The tools used in quantification can be based on alignment or without alignment. Alignment-based tools map all reads of a sample, to a genome or to transcriptome. Subsequently, quantify the reads that are assigned to a transcript, in the case of TopHat and RSEM [43, 45]. Tools that skip sequence alignment like HTSEq and featureCounts [48, 49], use the k-mer count, that is, they count all the k-mer in a sequencing library without aligning them to any reference, in this way the k-mer are counted and the unique k-mer are selected to quantify the expression and finally, these unique k-mer are assigned to the transcriptome to identify the transcription. (d) *Differential Expression Analysis*: At this stage, it is analyzed if the expression of a gene is different between different conditions. To determine if in a specific gene there are significant differences in the number of mapped reads corresponding to that gene, there are a large number of tools that are based on the comparison of the reading count for each transcript/gene under different biological conditions, by statistical analysis, which implies normalization methods since transcripts are synthesized at different levels (genes or transcripts with low or high level of expression), probabilistic models, modeling of reading counts at given distribution etc. In the differential expression analysis by RNA-seq, should be considered that the longer transcripts generate more reads compared to shorter transcripts. In addition, the technical noise introduced into the data during the sequencing process, as part of the variability in the number of reads produced by execution causes fluctuations in the number of mapped elements in the sample. To reduce the technical noise introduced into the data during the sequencing process, the number of reads must be normalized in order to obtain significant estimates of the expression. Among the statistical parameters used for this process are the metric of reads per kilobase per million mapped reads (RPKM), fragments per kilobase per million mapped reads (FPKM) [50, 51]. With these parameters it is possible to quantify transcription levels and make the comparison between samples. On the other hand, fold change allows us to evaluate the rate of change of a transcript in both conditions [52]. Within the challenges of transcriptome analysis, it is important to understand how the levels of expression differ in each situation studied, to achieve this objective, different methods try to model the biological variability such as EdgeR, DESeq, Cuffdiff [48, 53, 54]. In this way, there are currently different computational tools suitable for the overall study of the transcriptome suitable for each stage of analysis and specialized for each type of transcript under study (**Table 1**).

3.2 Bioinformatics tools in the study of coding RNA, non-coding RNA and microRNAs

The identification of non-coding RNAs and small RNAs is a vital issue in genetic analysis [29], in this sense algorithms have been developed for the analysis of this type of RNAs in particular (**Table 1**). Currently, the tools used to classify

Process	Tools	Objective	References
Quality analysis of reads	FastQC Fastx-toolkit Trimmomatic, PRINSEq, Flexbar	It analyzes the quality of the reads It debugs poor quality reads	[38–42]
Assembly	Trinity, Trans-ABySS, Oases, IDBA-Tran TOPHAT, STAR, IDBA-Tran, HISAT	Assembly of reads without genome or reference transcriptome Assembly of reads with genome or reference transcriptome	[46, 55, 56] [43, 44, 57, 58]
Classification of transcripts	BLAST, BLAT, GMAT, AUGUSTUS CPAT, FEELnc, NRC, lncRScan-SVM	It identifies coding transcripts by homology or by known transcript characteristics	[5, 59–61] [62–65]
Mapping	TOPHAT, STAR, HISAT, HISAT2, Bowtie	It aligns reads with a reference genome or transcriptome	[43, 44, 58, 66]
Quantification	RSEM, Feature Count StringTie, Salmon, Kallisto	It estimates the number of transcripts with or without their alignment	[45, 49, 67–69]
Classification of coding and non-coding transcripts	BEDTools, glbase	It determines the coordinates of the reference genome	[70]
	BLAST, BLAT, GMAP, AUGUSTUS	Through homology it manages to determine known sequences of transcripts found in databases	[5, 59–61]
	CPAT, FELLnc, lncRScan-SVM, NRC	It evaluates characteristics of coding and non-coding transcripts	[62–65]
Small RNA analysis	miRDeep Pic Tar	It quantifies known micro RNAs and identify new RNAs	[71–73]
	PiPMir	It identifies new micros RNAs in plants	[74]
	DARIO	It allows the recognition of micro RNAs, snoRNA and tRNA	[73]
	IntaRNA	It analyzes micro RNAs in eukaryotes and small bacterial RNAs	[75, 76]
	CopraRNA	It makes comparative predictions that include functional enrichment analysis	[76, 77]

Table 1.
Computational tools in the study of the transcriptome.

coding and non-coding sequences have two aspects, those that classify transcripts according to similarity and those that use known coding and non-coding properties [47]. Similarity-based tools classify transcripts, taking as reference the amino acid sequences of their transcripts translated with known protein coding genes, for example BLAST [5], BLATS [59], GMAP [60]. On the other hand, tools focused on coding and non-coding characteristics are based on the properties of known transcripts to predict whether a transcript encodes or not for a protein. The coding potential can be estimated using automatic learning approaches such as CPAT [62], FEELnc [63], lncRScan-SVM [64] and NRC [65]. These exclude transcripts based on properties such as transcription length, length of open reading frame (ORF), ORF coverage, k-mer frequency, codon usage bias, in addition to being optimized for different techniques [47]. In the choice of the tool to be used to evaluate the coding potential of a transcript, it will depend on what is sought in the study, if there is a good annotation and reference genome the tools based on similarity are practical and feasible in the analysis. However, in organisms that lack good gene annotations it is advisable to use tools based on coding and non-coding characteristics, which also allow to identify new genes. On the other hand, the availability of small readings opened a new field of study for small RNAs such as microRNAs (miRNAs), small RNAs of interference (siRNA) and piwiRNAs (piRNAs); Currently there are specialized tools for this type of RNA that provide additional biological knowledge. In this case miRDeep and its varieties are widely used to quantify known and novel RNA (miRNA), from the sequencing of small RNA by RNAseq [71, 72]; PiPMir [74] has been used for the detection of miRNA in plants. DARIO (http://dario.bioinf.uni-leipzig.de/index.py) is a web service that allows not only the recognition of new microRNAs but also small RNAs derived from other types of parental RNAs, such as snoRNA and tRNA [73]. Pic Tar is an algorithm for the identification of micro RNAs, which is based on functional interactions of micro RNA [78, 79]. IntaRNA has been designed for the study of micro RNAs in eukaryotes and small bacterial RNAs (RNAs) [75, 76]. CopraRNA is a comparative prediction algorithm that is complemented by post-processing methods that includes functional enrichment analysis [76, 77]. Finally, after analyzing the data, the biological conclusions must be carefully interpreted.

4. Proteomics data analysis

Transcriptome sequences provide resources for gene expression profile studies, as well as for the identification of mutations, sequence aberrations and RNA editing events [25], the above is possible to the existence of the open reading frame (ORF), however, in genomic data this does not imply the existence of a functional gene; despite the great advances in bioinformatics that facilitate the analysis and prediction of genes with the help of comparative genomics, and although they are years of development of molecular simulation methods, attempts to improve models that are already relatively close to the structure native, they have had little success, which may be due to inaccuracies in the potential functions used in simulations, such as the treatment of electrostatic and solvation effects or it may be necessary to improve sampling strategies due to the relatively long folding time scale of proteins; the combination of chemistry and physics with the large amount of information in known protein structures could provide a better route for the development of enhanced potential functions. Currently, it is difficult to accurately predict protein structures from genes, the success rate for the correct prediction of structures remains low [25, 80, 81]. Proteomics involves various technologies for deep proteome analysis, thus achieving quantification and identification of these proteins;

covering the part of functional analysis of genetic products, interaction studies, and protein localization, which helps explain the identity of an organism's proteins to know the structure and function. However, considering that the proteome is highly dynamic due to the complex regulatory systems that control the levels of protein expression, its use is limited, since in addition to the use of specialized personnel, facilities and equipment, software is also included for equipment, and databases, which increases costs [80, 82, 83]. Proteomics is constantly updated, generating challenges ranging from sample preparation to data collection. A large amount of information is generated from protein folding models, three-dimensional structures, prediction of unknown protein structures and functions, data obtained from the separation of proteins by electrophoresis in two-dimensional gels, isoelectric focusing, 2D protein visualization, peptide mass fingerprinting (PMF), MS, MS in tandem, etc., the above generates high performance proteomes with the help of bioinformatics, which introduces new algorithms to handle a large amount of heterogeneous data [84–86].

Some of the most used platforms in proteomics are: The Basic Local Alignment Search Tool (BLAST), Expert Protein Analysis System (ExPASy) and Protein Data Bank (PDB); BLAST (https://blast.ncbi.nlm.nih.gov/Blast.cgi). It is one of the most used and updated platforms, which uses simple but powerful methods for protein analysis comparing amino acid sequences, which makes it possible to determine homology between proteins, where the algorithms used to perform this procedure guarantee the best possible alignment, however, it does not guarantee the best structure [5, 86–90]. ExPASy gives access to a wide variety of databases and analytical tools dedicated to proteins and proteomics. On the other hand, PDB (https://www.wwpdb.org/) is the global repository of three-dimensional structures of macromolecules that is updated weekly and contains more than 153,000 protein structures, resulting from crystallographic studies, X-rays or nuclear magnetic resonance (NMR) created by modeling software, all these platforms contain various servers that help classify proteins according to their sequence, structure and function [86, 91, 92].

All this information is of great help, since it is used in different research areas, such as detection of diagnostic markers, candidates for vaccine production, understanding the mechanisms of pathogenicity, alteration of expression patterns in response to different signals and interpretation of functional protein pathways in different diseases [93–98].

5. Comparative genomics

Comparative genomics is a broad field of study that identifies differences between genomes and elucidates which of them are responsible for phenotypic changes in organisms [99]. In contrast to 'traditional' genomic studies that focus on a single genome per study [100], comparative genomics provides additional detailed information to that obtained from the analysis of a single genome, which can reveal the encoded functional potential of an organism compared to another [101–103]. Comparisons between different genomes of organisms lead to more rapid identification of different underlying mechanisms are shared between organisms and others that are different among them [104–106]. Likewise, comparative genomics allows a better understanding of how species have evolved [107]. In this sense, the concept of pangenome (**Figure 1**) refers to the set of genes in a particular species [106]. The commonly used partition of a pangenome considers three main parts: the central genome, the expendable or accessory genome and the singleton genome [108]. The central genes are responsible for the basic aspects of the biology of the species and its main phenotypic features; while accessory genes and singletons generally belong to supplementary biochemical pathways and functions that can confer selective advantages such as ecological adaptation [108]. While the global analysis of gene content (as in pangenome studies) provides information on differences in functional potential and possible phenotypic differences between organisms, specific central gene analyzes have also been used for studies of phylogenetic diversity [99, 108].

Initially, the concept of pangenome was used to refer to bacterial genomes, however, over time it has been used to refer to genomes of eukaryotic organisms such as yeasts [106, 109], plants [108, 110, 111], and viruses [108, 112]. Different organisms can be compared despite their phenotypic differences and with respect to their relationship of kinship (phylogenetic distances) [105, 113]. The assembly of genomes from sequencing data by Illumina or PacBio methods [114] involves five important stages, these steps are described in **Figure 2**, as well as some of the tools used [106].

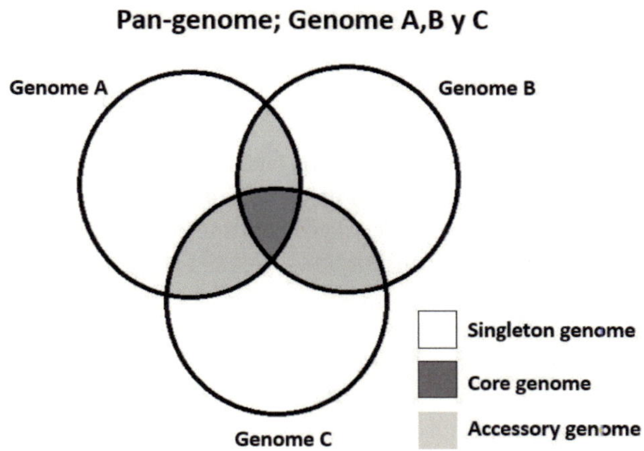

Figure 1.
Pangenome diagram of three different genomes.

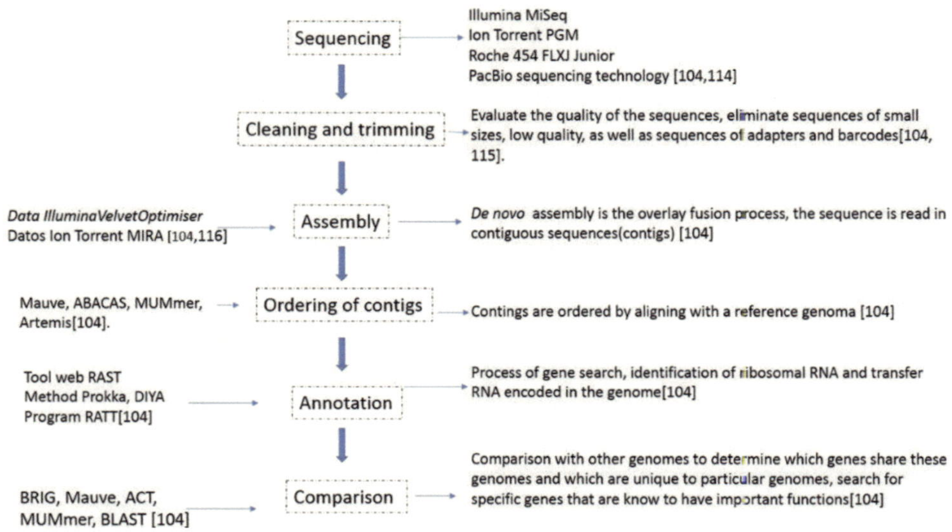

Figure 2.
Workflow for the de novo genome comparative analysis.

For gene comparisons databases with different characteristics are used, for example, to obtain gene families and identify their orthology the EDGAR database [108, 115] is used, as well as, the prokaryotic-genome analysis tool (PGAT) for the analysis of bacterial genomes [108, 116]. There are independent applications such as the Pan-genome analysis pipeline (PGAP) that have specific modules to perform the functional analysis of genes, the analysis and determination of each of the components of the pangenome, the detection of genetic variation as well as the analysis of Species evolution [108, 117], PanFunPro is a tool that allows pangenome analysis in protein prediction from genetic information [96]. There are tools that allow you to work with large amounts of data such as PanGP [118] and the large acale BSR [119].

The bacterial pan genome analysis tool (BPGA) [120] is a recently published package for pangenome analysis with seven functional modules; In addition to routine analysis, it presents a series of novel features for subsequent analyzes such as phylogeny, as well as tools that allow determining the presence and absence of certain genes in specific strains, another module to perform subset analysis, content analysis atypical G+C and KEGG & COG mapping of central, accessory and unique genes [108, 121–124].

6. Functional genomics

Functional genomics studies and assigns functions to the genome of an organism, including genes and non-genetic elements [125, 126], with the support of molecular and cellular biology studies, focused on the dynamic aspects of transcriptomics, proteomics and metabolomics [127], that allow to know the relationship of genes, their transcription, translation and protein-protein interactions [128, 129], that promote the phenotypic characteristics of each organism [125, 126]. A functional genomic approach can use multiple techniques for data analysis in a single study [129]. Apart from the tools of transcriptomics and proteomics, functional genomics needs of studies that allow us to know gene interactions [130, 131], genetic variations (polymorphisms) in different individuals through the study of SNPs [126, 132]. Likewise, it is important to know the regulation of genes in the expression of proteins that first carries out the analysis of promoter sequences, followed by the expression of the promoters and subsequently the expression of proteins [126, 133, 134]. Another study used for a rapid and systematic analysis of the expression of a large number of genes is the microarrays, which make it easier to observe the differential expression of genes from DNA or cDNA, as well as, allowing the finding gene functions novel and unexpected [135]. In addition, compare the pattern of gene expression under different conditions [136]. SAGE serial analysis of gene expression based on the study of cDNA allows to examine gene expression in a cell [126]. To perform a functional genomic observation, an assembled and identified genome must be had, which does not contain gaps, to avoid erroneous annotations. Subsequently, the assembled genome is compared with a reference genome, which together allows to predict genes. Next, the mapped elements are combined, and the biological information that allows to define an optimal set of annotations or functions is assigned. At the end, the data will have to be validated, this is achieved through manual inspections, experimental checks and quality measures [137]. To perform the genome annotation there are computational tools, one of the most used and friendly is Blast2GO which is a bioinformatics platform for high quality functional annotations and analysis of genomic data sets [138]. The data obtained can be shared with the public through databases so that other researchers can access them. Currently, GEO of NCBI is the public functional genomics database

Reference organisms	Databases	References
Escherichia coli	https://www.genome.jp/kegg-bin/show_organism?org=eco	[141]
Saccharomyces cerevisiae	https://www.yeastgenome.org/	[142]
Arabidopsis thaliana	https://www.arabidopsis.org/	[143]
Caenorhabditis elegans	https://wormbase.org/#012-34-5	[144]
Drosophila melanogaster	http://www.flybase.org/	[145]
Danio rerio	http://zfin.org/	[146]
Mus musculus	http://www.informatics.jax.org/	[147]
Homo sapiens: variation in humans	https://www.genome.jp/kegg-bin/show_organism?org=hsa	[148]

Table 2.
Databases of reference organisms used for genomic analysis.

that provides tools that help users in the consultation and download of data [139]. Likewise, KEGG is a database that is used as a tool to understand the high-level functions and utilities of the biological system, such as the cell, the organism or the ecosystem, based on molecular level information, generated by sequencing of the genome and other high performance [140]. There are also databases that store specific information on each of the most important model organisms (**Table 2**).

7. Phylogeny in the protein evolutionary process

The sequencing of the genome of an organism, has allowed to know the set of all its genes, elucidating the functions and products that they express, as well as the mechanisms of regulation in different metabolic processes, where endless proteins participate. To determine their possible functions, biochemical and genetic analyzes are used in a classical way, however, sequencing has contributed to the knowledge about the type of amino acids that make it up, and through the use of software multiple sequences have been aligned, where they have those that have been fully characterized as well as proteins where their biochemical characteristics are unknown and by homology between amino acids can be inferred in the functions that these proteins can present [149]. The use of bioinformatics, in protein analysis is a challenge, in recent years, phylogenetic profiles have been fundamental to relate homologous proteins by aligning their sequences, where it has been revealed that many share highly conserved regions and similar structures [150]. Phylogeny analyzes the changes that occur within the sequences and groups them in a diagram with ramifications, called a phylogenetic tree, all those sequences that belong to the same family can be grouped into a clade and in turn into subfamilies, providing data on their evolution and functional diversity [151].

Eukaryotic cells during their evolution have captured microorganisms that originated mitochondria, chloroplasts and other organelles, where their genes have been transferred to the nuclear genome, allowing the transport of encoded proteins in the nucleus. The different locations of proteins in the cell, and the different proteins that participate in cellular processes, have originated phylogenetic analyzes on the location of proteins in the cell, finding that they are closely related to prokaryotic proteins that have eukaryotes. The proteins of chloroplasts and mitochondria have a composition of amino acids, length, sequences and conserved regions very similar to those of prokaryotes [152, 153]. One of the limitations to analyze proteins among related organisms is that genomes must be complete, in order to determine the presence or absence of genes in these species [154].

The high number of sequences that are stored in the different databases, have allowed to infer in the evolutionary relationships of different proteins, which when presenting homology retain their function during long evolutionary times, however, homologous proteins can perform the same activity, but the substrates they use can come from different routes [155]. When organisms adapt to different environmental conditions they cause mutational changes in genome sequences, causing amino acid substitutions in enzymes, making them improve their efficiency and specificity, to maintain their catalytic function. Not all genes that code for proteins are susceptible to mutation, due to the presence of essential amino acids in function, stability and folding, and therefore a restriction is generated. Many of the mutations are usually random and, in those proteins, where these changes have been observed, it is due to an evolutionary pressure. If the protein plays an important role in the functions of the organism and the mutation brings improvements in activity, the change in the genome is maintained and optimized, favored by selective pressure, otherwise, when the function of the protein is not relevant. In the cell, the mutant gene is removed from the genome by random deletions. Evolutionary mechanisms have given rise to homologous protein families, which share a common ancestor [155]. The study of ancestral enzymes has suggested that these presented a high thermostability, due to the Precambrian era that was thermophilic, in addition to the fact that most microorganisms and other organisms adapted to these environments with high temperatures. The ancestral protein alignments with the current ones show evidence of a slow evolution in structure, but not in amino acids [156]. Therefore, enzymes are the product of years of evolution, where they have undergone changes to obtain a specific function, as well as greater affinity with the substrate and/or act on multi-substrates. Therefore, the genetic variability has generated homologous genes (they descend from a common ancestor and are called orthologs) that encode adapted proteins to perform their catalysis in extreme conditions. However, there are also paralogous genes, which have diverged, to encode proteins with different activities [157], many times a particular characteristic is preserved, such as the binding of a molecule or reaction mechanism, but they specialize in carrying out the same reaction but on different substrates, different regulation mechanisms, as well as cell localization. On the other hand, orthologous proteins tend to have the same function and their sequences have a high conservation [155].

To analyze these changes in the sequences, bioinformatics programs use algorithms and mathematical models, based on empirical matrices of amino acid substitution, as well as those that incorporate structural properties of the native state, such as secondary structure and accessibility [158]. Protein phylogeny studies are currently necessary to know protein-protein interactions in biological systems. Molecular or structural analyzes on proteins will require more information to respond if a protein is present in one or several species, as well as to predict the common ancestor and evolution times [159]. There are different methods to estimate the genetic distance of proteins, among the most used are the minimum distance, which predicts the phylogenetic relationship minimizing the total distance of the pairs of sequences adjacent nodes tree. While those of maximum parsimony and maximum likelihood, use the multiple sequence alignment, however, the maximum parsimony maximum builds a tree minimizing the total evolutionary changes between adjacent proteins and the maximum likelihood tries to minimize the probability of making such changes. The bioinformatics tools that use these algorithms are: TOPAL, Hennig86 and PAML, the computational packages that allow to occupy any of these are PHYLIP and PAUP, as well as MOLPHY, PASSML, PUZZLE, TAAR [160].

8. Protein modeling

One of the challenges of protein engineering and biology is to improve industrial processes, to achieve this it is necessary to determine the tertiary structure of proteins from the amino acid sequence, in order to design new proteins and even new medicines. Many of the protein structures that we know today have been obtained through experimentation by X-ray crystallography, NMR spectroscopy or cryo-EM, however, the large amount of proteins, makes these processes require more time and increase costs [161]. Modeling through bioinformatics programs has managed to predict the atomic structure of several proteins from their amino acid sequence, by comparison with known protein structures, commonly called templates, although these do not present an accuracy with traditional techniques, the processes are faster and more economical in addition to providing low resolution data during sequence comparison [162, 163]. If the protein studied presents a homolog of known structure, the analysis is easier and the generated model is of higher resolution, but if the homologs do not exist or are not identified, the modeling is constructed from scratch [164]. De novo modeling is based on the assembly of proteins using short peptide fragments, originating from known proteins based on similarity, although advances have been made using this process, it has only worked on proteins that contain less than 100 amino acids, on large proteins size is difficult to analyze due to lack of information, as well as the type of software used [161, 165].

The 3D protein structures provide data at the molecular level, functions and properties, among which are the study of the catalytic mechanism, design and improvement of ligands, union of macromolecules with proteins, functional relationships through structural similarity and identification of conserved residues [55]. The interest in finding new protein models is generating a large amount of data, which is being stored in different databases, including Protein Data Bank, where the coordinates of the experimentally obtained atoms are stored; until 2014 this base contained more than 80 million sequences and more than 100,000 experimentally obtained 3D structures [166, 167]. These data have allowed the classification of proteins in different hierarchical levels as family, superfamily and fold in relation to their structure and evolution. All those that are grouped into a family are evolutionarily related to high sequence similarity. It is suggested that the different families that maintain a structure and function, present a common ancestor and are grouped into superfamilies and the difference between these is due to the folds or secondary structure that they possess [160]. In the last decade, the predictions by computational models have revealed the structure and function of many proteins, but the advances have been in some cases slow and expensive, due to the programming methods used and the precision of these during modeling. Currently working on automated bioinformatics servers that will generate models with a high percentage of accuracy [168, 169]. One of the most used servers worldwide is SWIIS-MODEL, which was the first to model proteins through homology, and in recent years has been automated allowing complex modeling, as well as the introduction of the modeling engines ProMod3 and QMEAN [167, 170, 171]. Most modeling algorithms use the following steps: (1) Identification of related structures, (2) template choice, (3) target sequence alignment with templates, (4) molding construction, (5) model evaluation. However, one of the limitations during homology protein modeling is the choice of model proteins or templates as well as alignments against the problem sequences [172, 173]. When the similarity of the sequences between the problem protein and that of the databases is low, the relationship and alignment can be improved if structural information is included during the analysis [166]. Advances in biocomputing have allowed the generation of tools for modeling proteins that are more reliable and easier to use, reducing time and cost in the analysis. However, it is necessary to carry out experimentation to confirm that the prediction is correct, in addition to improving the efficiency of the techniques and with more known protein sequences and stored in the databases, therefore the different bioinformatics tools will play an important role in the postgenomic era [160].

9. Conclusions

Bioinformatics has evolved with daily work, which has allowed us to know how the biological molecules of a cell interact for their proper functioning, in addition to predicting various biological phenomena. In the last decade, the omic sciences have generated a great amount of data increasing the knowledge of the biological functions so that in the future they are able to predict diseases or formulate drugs with greater efficiency, however it is still necessary, to have a higher percentage of sequenced genes of the different organisms, as well as protein sequences, that allow enriching the databases, and with this more precise mathematical models are generated, which will benefit the computer programs so that they are more efficient, reliable, easy to use, reducing time and cost in the analyzes. This discipline becomes an essential part of biological studies every day, so its expansion and growth will be infinite, due to the evolutionary changes that are taking place in the cells caused by the different environmental phenomena.

Conflict of interest

The authors declare no conflict of interest.

Author details

Edna María Hernández-Domínguez[1], Laura Sofía Castillo-Ortega[1], Yarely García-Esquivel[1], Virginia Mandujano-González[2], Gerardo Díaz-Godínez[3] and Jorge Álvarez-Cervantes[1*]

1 Universidad Politécnica de Pachuca, Mexico

2 Universidad Tecnológica de la Corregidora, QRO, Mexico

3 Centro de Investigación en Ciencias Biológicas, Universidad Autónoma de Tlaxcala, Ixtacuixtla, Tlaxcala, Mexico

*Address all correspondence to: jorge_ac85@upp.edu.mx

References

[1] Benitez A, Cárdenas S. Bioinfomática en Colombia: Presente y futuro de la investigación biocomputacional. Biomédica. 2010;**3**:170-177. DOI: 10.7705/biomedica.v30i2.180

[2] Bernstein FC, Koetzle TF, Williams GJ, Meyer EF Jr, Brice MD, Rodgers JR. The protein data bank: A computer-based archival file for macromolecular structures. Journal of Molecular Biology. 1977;**112**:535-542. DOI: 10.1111/j.1432-1033.1977.tb11885.x

[3] Smith TF, Waterman MS. Identification of common molecular subsequences. Journal of Molecular Biology. 1981;**147**:195-197. DOI: 10.1016/0022-2836(81)90087-5

[4] Lipman DJ, Pearson WR. Rapid and sensitive protein similarity searches. Science. 1985;**227**:1435-1441. DOI: 10.1126/science.2983426

[5] Altschul SF, Gish W, Miller W, Myers EW, Lipman DJ. Basic local alignment search tool. Journal of Molecular Biology. 1990;**215**:403-410. DOI: 10.1016/S0022-2836(05)80360-2

[6] Meneses-Escobar CA, Rozo Murillo LV, Franco SJ. Tecnologías bioinformáticas para el análisis de secuencias de ADN. Scienctia et Technica. 2011;**16**:116-121

[7] Bustos RLS, Moreno LRD, Néstor D. Modelo de una bodega de datos para el soporte a la investigación bioinformática. Scientia et Technica. 2011;**16**:145-152

[8] Quíceno AHV. Bioinformática un Campo por conocer. Revista Electrónica de Veterinaria. 2006;**7**:1-9

[9] Harjinder SG, Prakash CR. Data Warehousing. La Integración de Información para la Mejor Toma de Decisiones. México: Prentice Hall; 1996. 382p

[10] Feng Z, Chen L, Maddula H, Akcan O, Oughtred R, Berman HM, et al. Ligand depot: A data warehouse for ligands bound to macrolecules. Bioinformatics. 2004;**20**:2153-2155. DOI: 10.1093/bioinformatics/bth214

[11] Judice LYK, Vladimir B. Database warehousing in bioinformatics. In: Bioinformatics Technologies. Berlin Heidelberg: Springer-Verlag; 2005. pp. 45-62. DOI: 10.1007/b138246

[12] Shualev V. Metabolomics technology and bioinformatics. Briefings in Bioinformatics. 2006;**7**:128-139. DOI: 10.1093/bib/bbl012

[13] Patti G, Yanes O, Siuzdak G. Metabolomics: The apogee of the omic triology. NIH Public Access. 2013;**13**:263-269. DOI: 10.1038/nrm3314

[14] Dalgliesh C, Horning E, Horning M, Knox K, Yarger K. A gas-liquid-chromatographic procedure for separating a wide range of metabolites occurring in urine or tissue extracts. The Biochemical Journal. 1966;**101**:792-810. DOI: 10.1038/nrm3314

[15] Horning E, Horning M. Metabolic profiles: Gas-phase methods for analysis of metabolites. Clinical Chemistry. 1971;**17**:802-809

[16] Ghezzi P, Floridi L, Boraschi D, Cuadrado A, Manda G, Levic S, et al. Oxidative stress and inflammation induced by environmental and psychological stressors: A biomarker perspective. Antioxidants & Redox Signaling. 2018;**20**:852-872. DOI: 10.1089/ars.2017.7147

[17] Kovatchev B. Diabetes technology: Markers, monitoring, assessment, and control of blood glucose fluctuations in diabetes. Scientifica (Cairo). 2012;**2012**:1-14. DOI: 10.6064/2012/283821

[18] Pourfarzam M, Zadhoush F. Newborn screening for inherited metabolic disorders; news and views. Journal of Research in Medical Sciences. 2013;**18**:801-808

[19] Jan S, Ahmad P. Ecometabolomics. Metabolic Fluxes versus Environmental Stoichiometry. Introducing Metabolomics. 1st ed. Cambridge: Academic Press; 2019. pp. 1-56

[20] Johnson C, Ivanisevic J, Benton H, Siuzdak G. Bioinformatics: The next frontier of metabolomics. Analytican Chemistry. 2015;**18**:801-808. DOI: 10.1021/ac5040693

[21] Johnson C, Patterson A, Idle J, González F. Xenobiotic metabolomics: Major impact on the metabolome. HHS Public Access. 2012;**52**:37-56. DOI: 10.1146/annurev-pharmtox-010611-134748

[22] Oliver S, Winson M, Kell D, Baganz F. Systematic functional analysis of the yeast genome. Trends in Biotechnology. 1998;**16**:373-378. DOI: 10.1016/S0167-7799(98)01214-1

[23] Kanehisa M, Goto S. KEGG: Kyptp encyclopedia of genes and genomes. Nucleic Acids Research. 2000;**28**:27-30. DOI: 10.1093/nar/28.1.27

[24] Caspi R, Billington R, Fulcher C, Keseler I, Kothari A, Krummenacker M, et al. The MateCyc database of metabolic pathways and enzymes. Nucleic Acids Research. 2018;**46**:D633-D339. DOI: 10.1093/nar/gkx935

[25] Morozova O, Hirst M, Marra MA. Applications of new sequencing technologies for transcriptome analysis. Annual Review of Genomics and Human Genetics. 2009;**10**:135-151. DOI: 10.1146/annurev-genom-082908-145957

[26] de Carvalho LM, Borelli G, Camargo AP, de Assis MA, Ferraz SMF, Fiamenghi MB, et al. Bioinformatics applied to biotechnology: A review towards bioenergy research. Biomass and Bioenergy. 2019;**123**:195-224. DOI: 10.1016/j.biombioe.2019.02.016

[27] Wang Z, Gerstein M, Snyder M. RNA-seq: A revolutionary tool for transcriptomics. Nature Reviews. Genetics. 2009;**10**:57. DOI: 10.1038/nrg2484

[28] Sedano JCS, Carrascal CEL. RNA-seq: herramienta transcriptómica útil para el estudio de interacciones planta-patógeno. Fitosanidad. 2012;**16**(2):101-113. DOI: 10.1093/bioinformatics/btr026

[29] Santana CIB. Buscando agujas en un pajar: viajes de RNAs pequenos in silico e in vitro. Acta Biológica Colombiana. 2011;**16**(3):103-113

[30] Peng M, Aguilar-Pontes MV, Hainaut M, Henrissat B, Hildén K, Mäkelä MR, et al. Comparative analysis of basidiomycete transcriptomes reveals a core set of expressed genes encoding plant biomass degrading enzymes. Fungal Genetics and Biology. 2018;**112**:40-46. DOI: 10.1016/j.fgb.2017.08.001

[31] Alwine JC, Kemp DJ, Stark GR. Method for detection of specific RNAs in agarose gels by transfer to diazobenzyloxymethyl-paper and hybridization with DNA probes. Proceedings of the National Academy of Sciences. 1977;**74**:5350-5354. DOI: 10.1073/pnas.74.12.5350

[32] Schena M, Shalon D, Davis RW, Brown PO. Quantitative monitoring of gene expression patterns with a complementary DNA microarray. Science. 1995;**270**(5235):467-470. DOI: 10.1126/science.270.5235.467

[33] Velculescu VE, Zhang L, Vogelstein B, Kinzler KW. Serial analysis of gene expression. Science.

1995;**270**:484-487. DOI: 10.1126/science.270.5235.484

[34] Seki M, Narusaka M, Kamiya A, Ishida J, Satou M, Sakurai T, et al. Functional annotation of a full-length Arabidopsis cDNA collection. Science. 2002;**296**:141-145. DOI: 10.1126/science.1071006

[35] Marguerat S, Bähler J. RNA-seq: From technology to biology. Cellular and molecular life sciences. Reino Unido. 2010;**67**:569-579. DOI: 10.1007/s00018-009-0180-6

[36] Parkinson J, Blaxter M. Expressed sequence tags. In: Parasite Genomics Protocols. Totowa: Humana Press; 2004. pp. 93-126. DOI: 10.1385/1-59259-793-9:075

[37] Nowrousian M. Next-generation sequencing techniques for eukaryotic microorganisms: Sequencing-based solutions to biological problems. Eukaryotic Cell. 2010;**9**:1300-1310. DOI: 10.1128/EC.00123-10

[38] Notes T, FAQ F. FastQC Tutorial & FAQ [Internet]. Rtsf.natsci.msu.edu. 2019. Available from: https://rtsf.natsci.msu.edu/genomics/tech-notes/fastqc-tutorial-and-faq/ [cited 30 August 2019]

[39] FASTX-Toolkit [Internet]. Bio.tools. 2019. Available from: https://bio.tools/fastx-toolkit [cited 30 August 2019]

[40] Bolger AM, Lohse M, Usadel B. Trimmomatic: A flexible trimmer for illumina sequence data. Bioinformatics. 2014;**30**:2114-2120. DOI: 10.1093/bioinformatics/btu170

[41] Schmieder R, Edwards R. Qality control and preprocessing of metagenomic datasets. Bioinformatics. 2011;**27**(6):863-864. DOI: 10.1093/bioinformatics/btr026

[42] Dodt M, Roehr J, Ahmed R, Dieterich C. FLEXBAR—Flexible barcode and adapter processing for next-generation sequencing platforms. Biology. 2012;**1**:895-905. DOI: 10.3390/biology1030895

[43] Strickler SR, Bombarely A, Mueller LA. Designing a transcriptome next-generation sequencing project for a nonmodel plant species. American Journal of Botany. 2012;**99**:257-266. DOI: 10.3732/ajb.1100292

[44] Dobin A, Davis CA, Schlesinger F, Drenkow J, Zaleski C, Jha S, et al. STAR: Ultrafast universal RNA-seq aligner. Bíoínformatícs (Oxford, England). 2013;**29**:15-21. DOI: 10.1093/bioinformatics/bts635

[45] Li B, Dewey CN. RSEM: Accurate transcript quantification from RNA-Seq data with or without a reference genome. BMC Bioinformatics. 2011;**12**:323. DOI: 10.1186/1471-2105-12-323

[46] Haas BJ, Papanicolaou A, Yassour M, Grabherr M, Blood PD, Bowden J, et al. De novo transcript sequence reconstruction from RNA-seq using the Trinity platform for reference generation and analysis. Nature Protocols. 2013;**8**:1494. DOI: 10.1038/nprot.2013.084

[47] Babarinde IA, Li Y, Hutchins AP. Computational methods for mapping, assembly and quantification for coding and non-coding transcripts. Computational and Structural Biotechnology Journal. 2019;**17**:628-637. DOI: 10.1016/j.csbj.2019.04.012

[48] Anders S, Huber W. Differential expression analysis for sequence count data. Genome Biology. 2010;**11**:R106. DOI: 10.1186/gb-2010-11-10-r106

[49] Liao Y, Smyth GK, Shi W. Feature counts: An efficient general-purpose program for assigning sequence reads to genomic features. Bioinformatics. 2013;**30**:923-930. DOI: 10.1093/bioinformatics/btt656

[50] Mortazavi A, Williams BA, McCue K, Schaeffer L, Wold B. Mapping and quantifying mammalian transcriptomes by RNA-seq. Nature Methods. 2008;**5**:621-628. DOI: 10.1038/nmeth

[51] Li B, Ruotti V, Stewart RM, Thomson JA, Dewey CN. RNA-seq gene expression estimation with read mapping uncertainty. Bioinformatics. 2009;**26**:493-500. DOI: 10.1093/bioinformatics/btp692

[52] Auer PL, Doerge RW. Statistical design and analysis of RNA sequencing data. Genetics. 2010;**185**:405-416. DOI: 10.1534/genetics.110.114983

[53] edgeR: Differential expression analysis of digital gene expression data [Internet]. 1st ed. 2008. Available from: http://chagall.med.cornell.edu/RNASEQcourse/edgeRUsersGuide-2018.pdf [cited 30 August 2019]

[54] Trapnell C, Hendrickson DG, Sauvageau M, Goff L, Rinn JL, Pachter L. Differential analysis of gene regulation at transcript resolution with RNA-seq. Nature Biotechnology. 2013;**31**:46. DOI: 10.1038/nbt.2450

[55] Robertson G, Schein J, Chiu R, Corbett R, Field M, Jackman SD, et al. De novo assembly and analysis of RNA-seq data. Nature Methods. 2010;**7**:909. DOI: 10.1093/gigascience/giz039

[56] Marcel H, Schulz Daniel R, Zerbino MV, Ewan B. Oases: Robust de novo RNA-seq assembly across the dynamic range of expression levels. Bioinformatics. 2012;**28**:1086-1092. DOI: 10.1093/bioinformatics/bts094

[57] Peng Y, Leung HC, Yiu SM, Lv MJ, Zhu XG, Chin FY. IDBA-tran: A more robust de novo de Bruijn graph assembler for transcriptomes with uneven expression levels. Bioinformatics. 2013;**29**:26-334. DOI: 10.1093/bioinformatics/btt219

[58] Kim D, Langmead B, Salzberg SL. HISAT: A fast-spliced aligner with low memory requirements. Nature Methods. 2015;**12**:357. DOI: 10.1038/nmeth.3317

[59] Kent WJ. BLAT—The BLAST-like alignment tool. Genome Research. 2002;**12**:656-664. DOI: 10.1101/gr.229202

[60] Wu TD, Watanabe CK. GMAP: A genomic mapping and alignment program for mRNA and EST sequences. Bioinformatics. 2005;**21**:1859-1875. DOI: 10.1093/bioinformatics/bti310

[61] Hoff KJ, Stanke M. Predicting genes in single genomes with augustus. Current Protocols in Bioinformatics. 2019;**65**:57. DOI: 10.1002/cpbi.57

[62] Wang L, Park HJ, Dasari S, Wang S, Kocher JP, Li W. CPAT: Coding-potential assessment tool using an alignment-free logistic regression model. Nucleic Acids Research. 2013;**41**:e74-e74. DOI: 10.1093/nar/gkt006

[63] Wucher V, Legeai F, Hedan B, Rizk G, Lagoutte L, Leeb T, et al. FEELnc: A tool for long non-coding RNA annotation and its application to the dog transcriptome. Nucleic Acids Research. 2017;**45**:e57-e57. DOI: 10.1093/nar/gkw1306

[64] Sun L, Liu H, Zhang L, Meng J. lncRScan-SVM: A tool for predicting long non-coding RNAs using support vector machine. PLoS One. 2015;**10**(10):e0139654. DOI: 10.1371/journal.pone.0139654

[65] Fiannaca A, La Rosa M, La Paglia L, Rizzo R, Urso A. nRC: Non-coding RNA classifier based on structural features. BioData Mining. 2017;**10**:1-27. DOI: 10.1186/s13040-017-0148-2

[66] Langmead B. Aligning short sequencing reads with bowtie. Current

Protocols in Bioinformatics. 2010;**32**:11-17. DOI: 10.1002/0471250953.bi1107s32

[67] Pertea M, Pertea GM, Antonescu CM, Chang TC, Mendell JT, Salzberg S. StringTie enables improved reconstruction of a transcriptome from RNA-seq reads. Nature Biotechnology. 2015;**33**:290. DOI: 10.1038/nbt.3122

[68] Patro R, Duggal G, Love MI, Irizarry RA, Kingsford C. Salmon provides fast and bias-aware quantification of transcript expression. Nature Methods. 2017;**14**:417. DOI: 10.1038/nmeth.4197

[69] Bray NL, Pimentel H, Melsted P, Pachter L. Near-optimal probabilistic RNA-seq quantification. Nature Biotechnology. 2016;**34**:525. DOI: 10.1038/nbt.3519

[70] Hutchins AP, Jauch R, Dyla M, Miranda-Saavedra D. A framework for combining, analyzing and displaying heterogeneous genomic and high-throughput sequencing data. Cell Regeneration. 2014;**3**:1-15. DOI: 10.1186/2045-9769-3-1

[71] Friedländer MR, Chen W, Adamidi C, Maaskola J, Einspanier R, Knespel S, et al. Discovering microRNAs from deep sequencing data using miRDeep. Nature Biotechnology. 2008;**26**:407. DOI: 10.1038/nbt1394

[72] An J, Lai J, Lehman ML, Nelson CC. miRDeep*: An integrated application tool for miRNA identification from RNA sequencing data. Nucleic Acids Research. 2012;**41**:727-737. DOI: 10.1093/nar/gks1187

[73] Fasold M, Langenberger D, Binder H, Stadler PF, Hoffmann S. DARIO: A ncRNA detection and analysis tool for next-generation sequencing experiments. Nucleic Acids Research. 2011;**39**:112-117. DOI: 10.1093/nar/gkr357

[74] Breakfield NW, Corcoran DL, Petricka JJ, Shen J, Sae-Seaw J, Rubio-Somoza I, et al. High-resolution experimental and computational profiling of tissue-specific known and novel miRNAs in Arabidopsis. Genome Research. 2012;**22**:163-176. DOI: 10.1101/gr.123547.111

[75] Busch A, Richter AS, Backofen R. IntaRNA: Efficient prediction of bacterial sRNA targets incorporating target site accessibility and seed regions. Bioinformatics. 2008;**24**:2849-2856. DOI: 10.1093/bioinformatics/btn544

[76] Wright PR, Georg J, Mann M, Sorescu DA, Richter AS, et al. CopraRNA and IntaRNA: Predicting small RNA targets, networks and interaction domains. Nucleic Acids Research. 2014;**42**:119-123. DOI: 10.1093/nar/gku359

[77] Wright PR, Richter AS, Papenfort K, Mann M, Vogel J, Hess WR, et al. Comparative genomics boosts target prediction for bacterial small RNAs. Proceedings of the National Academy of Sciences. 2013;**110**:487-496. DOI: 10.1073/pnas.1303248110

[78] Krek A, Grün D, Poy MN, Wolf R, Rosenberg L, Epstein EJ, et al. Combinatorial microRNA target predictions. Nature Genetics. 2005;**37**:495. DOI: 10.1038/ng1536

[79] Lall S, Grün D, Krek A, Chen K, Wang YL, Dewey CN, et al. A genome-wide map of conserved microRNA targets in *C. elegans*. Current Biology. 2006;**16**:460-471. DOI: 10.1016/j.cub.2006.01.050

[80] Pandey A, Mann M. Proteomics to study genes and genomes. Nature. 2000;**405**:837-846. DOI: 10.1038/35015709

[81] Baker D, Sali A. Protein structure prediction and structural genomics. Science. 2001;**294**:93-96. DOI: 10.1126/science.1065659

[82] Seaton D, Graf K, Baerenfaller M, Stitt A, Millar A, Gruissem W. Photoperioric control of the Arabidopsis proteome reveals a translational coincidence mechanism. Molecular Systems Biology. 2018;**14**:e7962. DOI: 10.15252/msb.20177962

[83] Yanovsky M, Kay S. Molecular basis of seasonal time measurement in Arabidopsis. Nature. 2002;**419**:308-312. DOI: 10.1038/nature00996

[84] Blueggel M, Chamrad D, Meyr H. Bioinformatics in proteomics. Current Pharmaceutical Biotechnology. 2004;**5**:79-88. DOI: 10.1201/9781420027524

[85] Schmidt A, Forne I, Imhof A. Bioinformatic analysis of proteomics data. BMC Systems Biology. 2014;**8**:1-7. DOI: 10.1186/1752-0509-8-S2-S3

[86] Popov I, Nenov A, Petrov P, Vassilev D. Bioinformatics in proteomics: A review on methods and algorithms. Biotechnology and Biotechnological Equipment. 2009;**23**:1115-1120. DOI: 10.1080/13102818.2009.10817624

[87] Smoot M, Guerlain S, Pearson W. Visualization of near-optimal sequence alignments. Bioinformatics. 2004;**20**:953-958. DOI: 10.1371/journal.pone.0178059

[88] Needleman S, Wunsch C. A general method applicable to the search for similarities in the amino acid sequence of two proteins. Journal of Molecular Biology. 1970;**48**:443-453. DOI: 10.1016/0022-2836(70)90057

[89] Barton G. Sequence alignment for molecular replacement. Acta Crystallographica. 2007;**64**:25-32. DOI: 10.1107/S0907444907046343

[90] Johnson M, Zaretskaya I, Raytselis Y, Merezhuj Y, McGinnis S, Madden T. NCBI BLAST: A better web interface. Nucleic Acids Research. 2008;**36**:5-9. DOI: 10.1093/nar/gkn201

[91] Gasteiger E, Gattiker A, Hoogland C, Ivanyi I, Appel R, Bairoch A. ExPASy: The proteomics server for in-depth protein knowledge and analysis. Nucleic Acids Research. 2003;**31**:3784-3788. DOI: 10.1093/nar/gkg563

[92] Rose P, Bojan B, Chunxiao B, Wolfgang B, Dimitris D, David G, et al. The RCSB protein data bank: Redesigned web site and web services. Nucleic Acids Research. 2011;**39**:392-401. DOI: 10.1093/nar/gkg1021

[93] Aslam B, Basit M, Nisar M, Khurshid M. Proteomics: Technologies and their applications. Journal of Chromatographic Science. 2017;**55**:182-196. DOI: 10.1093/chromsci/bmw167

[94] Stroggilos R, Mokou M, Latosinska A, Makridakis M, Lygirou V, Mavrogeorgis E, et al. Proteome-based classification of non-muscle invasive bladder cancer. International Journal of Cancer. 2019. DOI: 10.1002/ijc.32556

[95] Chaudhary H, Nameirakpam J, Kumrah R, Pandiarajan V, Suri D, Rawat A, et al. Biomarkers for Kawasaki disease: Clinical utility and the challenges ahead. Frontiers in Pediatrics. 2019;**7**:1-10. DOI: 10.3389/fped.2019.00242

[96] Yatoo M, Parray R, Bhat R, Nazir Q, Haq A, Malik U, et al. Novel candidates for vaccine development against *Mycoplasma capricolum* subspecies Capripneumoniae (Mccp)—Current knowledge and future prospects. Vaccine. 2019;**7**:2-21. DOI: 10.3390/vaccines703007

[97] Burgos-Canul Y, Canto-Canché B, Berezovski M, Mironov G, Loyola-Vargas V, Barba de Rosa A, et al. The cell wall proteome from two

strains of *Pseudocercospora fijiensis* with differences in virulence. World Journal of Microbiology and Biotechnology. 2019;**35**:105. DOI: 10.1007/s11274-019-2681-2

[98] Parolo S, Marchetti L, Lauria M, Misselbeck K, Scott-Boyer M, Caberlotto L, et al. Combined use of protein biomarkers and network analysis unveils deregulated regulatory circuits in Duchenne muscular dystrophy. PLoS One. 2018;**13**:e0194225. DOI: 10.1371/journal.pone.0194225

[99] Hu B, Xie G, Lo C, Starkenburg SR, Chain PSG. Pathogen comparative genomics in the next-generation sequencing era: Genome alignments, pangenomics and metagenomics. Briefings in Functional Genomics. 2011;**10**:322-333. DOI: 10.1093/bfgp/elr042

[100] Ashburner M, Ball CA, Blake JA, Botstein D, Butler H, Cherry JM, et al. Gene ontology: Tool for the unification of biology. Nature Genetics. 2000;**25**:25-29. DOI: 10.1038/75556

[101] Mira A, Martín-Cuadrado AB, D'Auria G, Rodríguez-Valera F. The bacterial pan-genome: A new paradigm in microbiology. International Microbiology. 2010;**13**:45-57. DOI: 10.2436/20.1501.01.110

[102] Loman NJ, Constantinidou C, Chan JZ, Halachev M, Sergeant M, Penn CW, et al. High-throughput bacterial genome sequencing: An embarrassment of choice, a world of opportunity. Nature Reviews. Microbiology. 2012;**10**:599-606. DOI: 10.1038/nrmicro2850

[103] Stahl PL, Lundeberg J. Toward the single-hour high-quality genome. Annual Review of Biochemistry. 2012;**81**:359-378. DOI: 10.1146/annurev-biochem-060410-094158

[104] Edwards DJ, Holt KE. Beginner's guide to comparative bacterial genome analysis using next-generation sequence data. Microbial Informatics and Experimentation. 2013;**3**:2. DOI: 10.1186/2042-5783-3-2

[105] Hardison RC. Comparative genomics. PLoS Biology. 2003;**1**:156-160. DOI: 10.1371/journal.pbio.0000058

[106] Tettelin H, Riley D, Cattuto C, Medini D. Comparative genomics: The bacterial pan-genome. Current Opinion in Microbiology. 2008;**11**:472-477. DOI: 10.1016/j.mib.2008.09.006

[107] Mosquera-Rendón J, Rada-Bravo AM, Cárdenas-Brito S, Corredor M, Restrepo-Pineda E, Benítez-Páez A. Pangenome-wide and molecular evolution analyses of the *Pseudomonas aeruginosa* species. BMC Genomics. 2016;**17**(45):1-14. DOI: 10.1186/s12864-016-2364-4

[108] Zekic T, Holley G, Stoye J. Pan-genome storage and analysis techniques. In: Setubal JC, Peter JS, Stadler F, editors. Comparative Genomics Methods and Protocols. Totowa: Humana Press; 2018. pp. 29-54. DOI: 10.1007/978-1-4939-7463-4. ch2

[109] Cliften PF, Hillier LW, Fulton L, Graves T, Miner T, Gish WR, et al. Surveying *Saccharomyces* genomes to identify functional elements by comparative DNA sequence analysis. Genome Research. 2001;**11**:1175-1186. DOI: 10.1101/gr.182901

[110] Hirsch CN, Foerster JM, Johnson JM, Sekhon RS, Muttoni G, Vaillancourt B, et al. Insights into the maize pan-genome and pan-transcriptome. The Plant Cell. 2014;**26**:121-135. DOI: 10.1105/tpc.113.119982

[111] Weigel D, Mott R. The 1001 genomes project for *Arabidopsis*

thaliana. Genome Biology. 2009;**10**:107. DOI: 10.1186/gb-2009-10-5-107

[112] Huang S, Zhang S, Jiao N, Chen F. Comparative genomic and phylogenomic analyses reveal a conserved core genome shared by estuarine and oceanic cyanopodoviruses. PLoS One. 2015;**10**: 1-17. DOI: 10.1371/journal.pone.0142962

[113] Rubin GM, Yandell MD, Wortman JR, Miklos GLG, Nelson CR, Hariharan IK, et al. Comparative genomics of the eukaryotes. Science. 2000;**287**:2204-2215. DOI: 10.1007/978-1-4939-7463-4_3

[114] Hassan YI, Lepp D, Zhou T. Next-generation whole-genome sequencing platforms and factors to consider for bacterial applications. Journal of Microbiology, Biotechnology and Food Sciences. 2015;**5**:29-33. DOI: 10.15414/jmbfs.2015.5.1.29-33

[115] Blom J, Kreis J, Sp€anig S, Juhre T, Bertelli C, Ernst C, et al. EDGAR 2.0: An enhanced software platform for comparative gene content analyses. Nucleic Acids Research. 2016;**44**:22-28. DOI: 10.1093/nar/gkw255

[116] Brittnacher MJ, Fong C, Hayden HS, Jacobs MA, Radey M, Rohmer L. PGAT: A multistrain analysis resource for microbial genomes. Bioinformatics. 2011;**27**:2429-2430. DOI: 10.1093/bioinformatics/btr418

[117] Zhao Y, Wu J, Yang J, Sun S, Xiao J, Yu J. PGAP: Pan-genomes analysis pipeline. Bioinformatics. 2012;**28**:416-418. DOI: 10.1093/bioinformatics/btr655

[118] Zhao Y, Jia X, Yang J, Ling Y, Zhang Z, Yu J, et al. PanGP: A tool for quickly analyzing bacterial pan-genome profile. Bioinformatics. 2014;**30**:1297-1299. DOI: 10.1093/bioinformatics/btu017

[119] Sahl JW, Gregory Caporaso J, Rasko DA, Keim P. The large-scale blast score ratio (LS-BSR) pipeline: A method to rapidly compare genetic content between bacterial genomes. PeerJ. 2014;**2**:e332. DOI: 10.7717/peerj.332

[120] Chaudhari NM, Gupta VK, Dutta C. BPGA-an ultra-fast pan-genome analysis pipeline. Scientific Reports. 2016;**6**:1-10. DOI: 10.1038/srep24373

[121] Galperin MY, Koonin EV. Comparative genome analysis. In: Baxevanis AD, Francis Ouellette BF, editors. Bioinformatics: A Practical Guide to the Analysis of Genes and Proteins. 2nd ed. Hoboken: John Wiley & Sons, Inc.; 2001. pp. 359-392. DOI: 10.1093/bib/bbk012

[122] Wattam AR, Thomas Brettin T, James J, Davis JJ, Svetlana Gerdes S, Kenyon R, et al. Assembly, annotation, and comparative genomics in PATRIC, the all bacterial bioinformatics resource center. In: Setubal JC, Peter JS, Stadler F, editors. Comparative Genomics Methods and Protocols. 1st ed. Totowa: Humana Press; 2018. pp. 79-102. DOI: 10.1007/978-1-4939-7463-4

[123] Santos AR, Barbosa E, Fiaux K, Zurita-Turk M, Chaitankar V, Kamapantula B, et al. PANNOTATOR: An automated tool for annotation of pan-genomes. Genetics and Molecular Research. 2013;**12**:2982-2989. DOI: 10.4238/2013

[124] Angiuoli SV, Hotopp JCD, Salzberg SL, Tettelin H. Improving pan-genome annotation using whole genome multiple alignment. BMC Bioinformatics. 2011;**12**:272-283. DOI: 10.1186/1471-2105-12-272

[125] Pevsner J. Bioinformatics and Functional Genomics. 3rd ed. Hoboken:

Wiley Blackwell; 2015. pp. 635-695. DOI: 10.1002/9780470451496

[126] Kaushik S, Sharma D. Functional genomics. Reference module in life sciences. Encyclopedia of Bioinformatics and Computational Biology. 2018. DOI: 10.1016/b978-0-12-809633-8.20222-7

[127] Bino RJ, Hall RD, Fiehn O, Kopka J, Saito K, Draper J, et al. Potential of metabolomics as a functional genomics tool. Trends in Plant Science. 2004;**9**:418-425. DOI: 10.1016/j.tplants.2004.07.004

[128] Miller W, Makova KD, Nekrutenko A, Hardison RC. Comparative genomics. Annual Review of Genomics and Human Genetics. 2004;**5**:15-56. DOI: 10.1146/annurev.genom.5.061903.180057

[129] Jones AR, Miller M, Aebersold R, Apweiler R, Ball CA, Brazma A, et al. The functional genomics experiment model (FuGE): An extensible framework for standards in functional genomics. Nature Biotechnology. 2007;**25**:1127-1133. DOI: 10.1038/nbt1347

[130] Schlitt T, Palin K, Rung J, Dietmann S, Lappe M, Ukkonen E, et al. From gene networks to gene function. Genome Research. 2003;**13**:2568-2576. DOI: 10.1101/gr.1111403

[131] Boucher B, Jenna S. Genetic interaction networks: Better understand to better predict. Frontiers in Genetics. 2013;**4**:1-16. DOI: 10.3389/fgene.2013.00290

[132] Karchin R. Next generation tools for the annotation of human SNPs. Briefings in Bioinformatics. 2009;**10**:35-52. DOI: 10.1093/bib/bbn047

[133] Zhu J, Zhang MQ. SCPD: A promoter database of the yeast *Saccharomyces cerevisiae*. Bioinformatics. 1999;**15**:607-611. DOI: 10.1093/bioinformatics/15.7.607

[134] Collado-Vides J, Salgado H, Morett E, Gama-Castro S, Jiménez-Jacinto V, Martínez-Flores I, et al. Bioinformatics resources for the study of gene regulation in bacteria. Journal of Bacteriology. 2009;**191**:23-31. DOI: 10.1128/JB.01017-08

[135] Slonim K, Yanai I. Getting started in gene expresión microarray analysis. PLoS Computational Biology. 2009;**5**:e1000543. DOI: 10.1371/journal.pcbi.1000543

[136] Miller MB, Tang YW. Basic concepts of microarrays and potential applications in clinical microbiology. Clinical Microbiology Reviews. 2009;**22**:611-633. DOI: 10.1128/CMR.00019-09

[137] Alvarado VJ. Anotación de genoma. Conogasi.org 2019. Sitio web: http://conogasi.org/articulos/anotacion-de-genoma/ [cited 18 August 2019]

[138] Conesa A, Götz S, García-Gómez JM, Terol J, Talón M, Robles M. Blast2GO: A universal tool for annotation, visualization and analysis in functional genomics research. Bioinformatics. 2005;**21**:3674-3676. DOI: 10.1093/bioinformatics/bti610

[139] Barrett T, Troup DB, Wilhite SE, Ledoux P, Rudnev D, Evangelista C, et al. NCBI GEO: Archive for high-throughput functional genomic data. Nucleic Acids Research. 2009;**37**:885-890. DOI: 10.1093/nar/gkn764

[140] KEGG: Kyoto Encyclopedia of Genes and Genomes. Available from: https://www.genome.jp/kegg/ [cited 17 August 2019]

[141] Brown SD, Jun S. Complete genome sequence of *Escherichia coli* NCM3722. Genome Announcements.

2015;**3**(4):00879-15. DOI: 10.1128/genomea.00879-15

[142] Saccharomyces genoma database. 2019. Available from: https://www.yeastgenome.org/ [17 August 2019]

[143] Tair Phoenix bioinformatics. 2019. Available from: https://www.arabidopsis.org [17 August 2019]

[144] WormBase versión: WS271. 2019. Available from: https://wormbase.org/#012-34-5 [17 August 2019]

[145] A Database of Drosophila Genes & Genomes. 2019. Available from: http://www.flybase.org [17 August 2019]

[146] The Zebrafish Information Network, University of Oregon. 2019. Available from: http://zfin.org/ [17 August 2019]

[147] Mouse Genome Informatics. The Jackson Laboratory. 2019. Available from: http://www.informatics.jax.org/. [17 August 2019]

[148] *Homo sapiens* (Human). 2019. Available from: https://www.genome.jp/kegg-bin/show_organism?org=hsa [17 August 2019]

[149] Marcotte EM, Pellegrini M, Ng HL, Rice DW, Yeates T, Eisenberg D. Detecting protein function and protein-protein interactions from genome sequences. Science. 1999;**285**:751-753. DOI: 10.1126/science.285.5428.751

[150] Pellegrini M, Marcotte EM, Thompson MJ, Eisenberg D, Yeates T. Assigning protein functions by comparative genome analysis: Protein phylogenetic profiles. Proceedings of the National Academy of Sciences. 1999;**96**:4285-4288. DOI: 10.1073/pnas.96.8.4285

[151] Song L, Wu S, Tsang A. Phylogenetic analysis of protein family. In: de Vries R, Tsang A, Grigoriev I, editors. Fungal Genomics. Methods in Molecular Biology. New York, NY: Humana Press; 2018. pp. 267-275. DOI: 10.1007/978-1-4939-7804-5

[152] Margulis L. Origin of Eukaryotic Cells: Evidence and Research Implications for a Theory of the Origin and Evolution of Microbial, Plant, and Animal Cells on the Precambrian Earth. New Haven: Yale University Press; 1970. p. 349

[153] Marcotte EM, Xenarios I, Van der Bliek AM, Eisenberg D. Localizing proteins in the cell from their phylogenetic profiles. Proceedings of the National Academy of Sciences. 2000;**97**:12115-12120. DOI: 10.1073/pnas.220399497

[154] Valencia A, Pazos F. Computational methods for the prediction of protein interactions. Current Opinion in Structural Biology. 2002;**12**:368-373. DOI: 10.1016/S0959-440X(02)00333-0

[155] Kaminska KH, Milanowska K, Bujnicki JM. The basics of protein sequence analysis. In: Bujnicki JM, editor. Prediction of Protein Structures, Functions, and Interactions. Hoboken: John Wiley & Sons, Ltd.; 2009. pp. 1-38. DOI: 10.1002/9780470741894

[156] Merkl R, Sterner R. Ancestral protein reconstruction: techniques and applications. Biological Chemistry. 2016;**397**:1-21. DOI: 10.1515/hsz-2015-0158

[157] Tyzack JD, Furnham N, Sillitoe I, Orengo CM, Thornton JM. Understanding enzyme function evolution from a computational perspective. Current Opinion in Structural Biology. 2017;**47**:131-139. DOI: 10.1016/j.sbi.2017.08.003

[158] Bastolla U, Arenas M. The influence of protein stability on sequence evolution: Applications to phylogenetic inference. In: Sikosek T, editor. Computational Methods in Protein Evolution. New York, NY: Humana Press; 2019. pp. 215-231. DOI: 10.1007/978-1-4939-8736-8_11

[159] Szurmant H, Weigt M. Inter-residue, inter-protein and inter-family coevolution: Bridging the scales. Current Opinion in Structural Biology. 2018;**50**:26-32. DOI: 10.1016/j.sbi.2017.10.014

[160] Xu D, Xu Y, Uberbacher CE. Computational tools for protein modeling. Current Protein & Peptide Science. 2000;**1**:1-21. DOI: 10.2174/1389203003381469

[161] Cheung NJ, Yu W. De novo protein structure prediction using ultra-fast molecular dynamics simulation. PLoS One. 2018;**13**:e0205819. DOI: 10.1371/journal.pone.0205819.

[162] Bonneau R, Baker D. Ab initio protein structure prediction: Progress and prospects. Annual Review of Biophysics and Biomolecular Structure. 2001;**30**:173-189. DOI: 10.1146/annurev.biophys.30.1.173

[163] Hung L, Ngan S, Samudrala R. De novo protein structure prediction. In: Xu Y, Xu D, Liang J, editors. Computational Methods for Protein Structure Prediction and Modeling. New York: Springer; 2007. pp. 43-64. DOI: 10.1007/978-0-387-68825-1_2

[164] Lee J, Freddolino PL, Zhang Y. Ab initio protein structure prediction. In: Rigden DJ, editor. From Protein Structure to Function with Bioinformatics. Dordrecht: Springer; 2017. pp. 3-35. DOI: 10.1007/978-94-024-1069-3_1

[165] Shen Y, Bax A. Homology modeling of larger proteins guided by chemical shifts. Nature Methods. 2015;**12**:747. DOI: 10.1038/nmeth.3437

[166] Kelleym LA, Mezulis S, Yates CM, Wass MN, Sternberg MJ. The phyre2 web portal for protein modeling, prediction and analysis. Nature Protocols. 2015;**10**:845. DOI: 10.1038/nprot.2015.053

[167] Waterhouse A, Bertoni M, Bienert S, Studer G, Tauriello G, Gumienny R, et al. SWISS-MODEL: Homology modelling of protein structures and complexes. Nucleic Acids Research. 2018;**46**:W296-W303. DOI: 10.1093/nar/gky427

[168] Kryshtafovych A, Barbato A, Fidelis K, Monastyrskyy B, Schwede T, Tramontano A. Assessment of the assessment: Evaluation of the model quality estimates in CASP10. Proteins. 2014;**82**:112-126. DOI: 10.1002/prot.24347

[169] Yang J, Zhang Y. Protein structure and function prediction using I-TASSER. Current Protocols in Bioinformatics. 2015;**52**:5-8. DOI: 10.1002/0471250953.bi0508s52

[170] Benkert P, Biasini M, Schwede T. Toward the estimation of the absolute quality of individual protein structure models. Bioinformatics. 2011;**27**:343-350. DOI: 10.1093/bioinformatics/btq662

[171] Biasini M, Schmidt T, Bienert S, Mariani V, Studer G, Haas J, et al. OpenStructure: An integrated software framework for computational structural biology. Acta Crystallographica, Section D: Biological Crystallography. 2013;**69**:701-709. DOI: 10.1107/S0907444913007051

[172] Fiser A, Šali A. Modeller: Generation and refinement of homology-based protein structure models. In: Methods in Enzymology. Cambridge: Academic Press; 2003. pp. 461-491. DOI: 10.1016/S0076-6879(03)74020-8

[173] Song Y, DiMaio F, Wang RYR, Kim D, Miles C, Brunette TJ, et al. High-resolution comparative modeling with RosettaCM. Structure. 2013;**21**:1735-1742. DOI: 10.1016/j.str.2013.08.005

Chapter 4

Scaffolding Contigs Using Multiple Reference Genomes

Yi-Kung Shieh, Shu-Cheng Liu and Chin Lung Lu

Abstract

Scaffolding is an important step of the genome assembly and its function is to order and orient the contigs in the assembly of a draft genome into larger scaffolds. Several single reference-based scaffolders have currently been proposed. However, a single reference genome may not be sufficient alone for a scaffolder to correctly scaffold a target draft genome, especially when the target genome and the reference genome have distant evolutionary relationship or some rearrangements. This motivates researchers to develop the so-called multiple reference-based scaffolders that can utilize multiple reference genomes, which may provide different but complementary types of scaffolding information, to scaffold the target draft genome. In this chapter, we will review some of the state-of-the-art multiple reference-based scaffolders, such as Ragout, MeDuSa and Multi-CAR, and give a complete introduction to Multi-CSAR, an improved extension of Multi-CAR.

Keywords: bioinformatics, sequencing, contig, scaffolding, multiple reference genomes

1. Introduction

Due to recent advances in next-generation sequencing (NGS) technologies, more and more genomes of organisms can be sequenced quickly at a moderate cost [1]. However, assembling a large number of reads generated from current NGS sequencing platforms into a complete genome still is a challenging job [2]. Largely because of repetitive sequences, whose lengths are often larger than those of the reads, most of assembled sequences are just *draft* genomes that usually consists of several hundreds or even thousands of *contigs* (contiguous sequences). The availability of complete genomes actually is significant to the downstream analysis and interpretation of their sequences in many biological applications [3]. To further obtain more complete sequences of draft genomes, therefore, the contigs of the draft genomes usually are required to be ordered and oriented into *scaffolds*, which actually are larger gap-containing sequences whose gaps between the scaffolded contigs can be closed later in the gap-filling process [4].

The scaffolding process utilizes a genomic sequence available from a related organism to serve as a *reference* to scaffold the contigs of a draft genome. So far, many such reference-based scaffolders have been proposed [5–14]. The algorithms used to develop all these scaffolders can be classified into two main categories: the *alignment-based* algorithms [5–10] and the *rearrangement-based* algorithms [11–14]. The alignment-based scaffolding algorithms first align contigs in a target draft genome against a reference sequence and then scaffold the contigs according to the positions of their matches in the reference. On the other hand, the rearrangement-based scaffolding algorithms utilize the concept of genome rearrangements to scaffold the contigs of the target draft genome such that the sequence markers (or genes) shared between the scaffolded target and reference genomes have similar order and orientation as much as possible.

In some cases, it may be insufficient for a scaffolder to utilize only one single genome as the reference for correctly computing the scaffolds of a target draft genome, in particular when the target and reference genomes have a distant phylogenetic relationship or they have undergone some kinds of rearrangements, such as reversals, transpositions, block-interchanges and translocations. This situation inspires the requirement for developing multiple reference-based scaffolders, expecting that they can refer to several different but complementary genomes to order and orient the contigs of the target genome.

2. State-of-the-art multiple reference-based scaffolders

Below, we review three state-of-the-art multiple reference-based scaffolders: Ragout [15], MeDuSa [16] and Multi-CAR [17].

2.1 Ragout

Ragout (Reference-Assisted Genome Ordering UTility) is a rearrangement-based scaffolder for ordering and orienting the contigs of a draft genome using multiple reference genomes [15]. The input of Ragout includes a target draft genome, multiple reference genomes, and a phylogenetic tree between them. Ragout uses different colors to display the target and reference genomes and further represents all of these genomes as sequences of *synteny blocks*. Ragout then creates a so-called *incomplete multi-color breakpoint graph*, in which vertices represent the ends of synteny blocks and edges denote adjacencies of two synteny blocks occurring in the target and reference genomes. For the purpose of distinction, the edges are also colored by Ragout using the colors of the corresponding genomes. Because the target genome is already fragmented into contigs, some adjacencies of synteny blocks in the target genome are missing. Ragout tries to recover these missing adjacencies by using other existing adjacencies from the reference genomes. In the recovery process, Ragout computes the parsimony costs of all possible missing adjacencies by solving a so-called *half-breakpoint state parsimony problem* on the given phylogenetic tree, which actually is an NP-hard (non-deterministic polynomial time-hard) problem, meaning that it is hard to compute its optimal solution in polynomial time. Therefore, a heuristic approach is applied by Ragout to calculate the approximate parsimony costs of all the missing adjacencies. A perfect matching with minimum cost is then computed by Ragout on a graph created by using the missing adjacencies and is further used to scaffold the contigs of the target genome. Actually, the above procedure is repeated by Ragout multiple times with using different sizes of synteny blocks and moreover the scaffolding results obtained from all these iterations are then combined into a single set of scaffolds. Finally, a refinement is performed by Ragout to insert a number of small but repetitive contigs back to the resulting scaffolds.

2.2 MeDuSa

MeDuSa (Multi-Draft based Scaffolder) is a multiple reference-based scaffolder that does not require a given phylogenetic tree for the target and references genomes [16]. From the given target and reference genomes, MeDuSa constructs a so-called *scaffolding graph*, which denotes by vertices the contigs of the target genome and by edges the adjacencies between any two contigs when they can be mapped to the reference genomes. Moreover, each edge in the scaffolding graph is associated with a *weight* to represent the number of reference genomes supporting the existence of the edge. As a result, it is not hard to see that a *path cover*, which is a set vertex-disjoint paths covering all the vertices of the scaffolding graph, denotes a set of scaffolds in the target genome. Unfortunately, however, finding a path cover of maximum weight in a graph is already known as an NP-hard problem. Therefore, MeDuSa utilizes a 2-approximation algorithm to find an approximate path cover from the scaffolding graph. Finally, MeDuSa applies a majority rule to determine the orientations of contigs on each path of the approximate path cover.

2.3 Multi-CAR

Multi-CAR (Multiple reference-based Contig Assembly using Rearrangements) is multiple-reference version of CAR (Contig Assembly using Rearrangements) [17]. CAR actually is a single reference-based scaffolder that utilizes a complete reference genome to scaffold the contigs of a target draft genome [13]. Like MeDuSa, Multi-CAR does not require prior knowledge concerning phylogenetic relationships among target and reference genomes. However, in contrast to Ragout and MeDuSa, both attempting to solve an NP-hard problem in their scaffolding processes, the algorithm behind Multi-CAR involves only polynomially solvable problems, as described as follows. First, Multi-CAR utilizes CAR to compute a single reference-derived scaffolding result for a target draft genome based on each of multiple reference genomes. Second, Multi-CAR uses all single reference-derived scaffolds to build an edge-weighted *contig adjacency graph*. In this contig adjacency graph, the vertices denote extremities of contigs (i.e., each contig is represented by two vertices) and the edges represent whether two contigs are ordered consecutively in a scaffold returned by CAR based on a single reference genome (if so, the adjacent extremities of these two contigs are connected by an edge). In addition, if there are multiple reference genomes to *support* an edge connection, then this edge will be assigned a weight that equals to the sum of the weights of the supporting reference genomes. The weight of each reference genome is given by the users in advance; otherwise, it is defaulted to one. Third, Multi-CAR continues to find a maximum weighted perfect matching from the contig adjacency graph. Finally, Multi-CAR constructs a multiple reference-derived scaffold for the target draft genome according to the maximum weighted perfect matching.

3. A recent multiple reference-based scaffolder

In this section, we give a detailed introduction to a recent multiple reference-based scaffolder, called Multi-CSAR (Multiple reference-based Contig Scaffolder using Algebraic Rearrangements), which is an improved extension of Multi-CAR [18]. Unlike Ragout and MeDuSa, Multi-CAR actually can not accept incomplete genomes as references, which greatly limits the widespread adoption of Multi-CAR because complete reference genomes are not always available for a target draft genome in practical usage [19]. In addition, the weight of all reference genomes used by Multi-CAR must be assigned by the users; otherwise, they are defaulted to one. However, it is usually not easy for the ordinary users to correctly determine these weights. Therefore, Multi-CSAR has been developed to further overcome these limitations of Multi-CAR. In principle, the main steps of the algorithm in

Multi-CSAR is the same as those in Multi-CAR, except that Multi-CSAR utilizes CSAR [14], instead of CAR [13], to compute the single reference-derived scaffolding result for the target draft genome, and also designs a *sequence identity-based weighting scheme* to automatically derive the weights of all the reference genomes. CSAR actually is an improved version of CAR and their main difference in usage is that the reference genome used by CAR needs to be complete, but the one used by CSAR can be incomplete.

3.1 Algorithm of multi-CSAR

Suppose that T denotes a target draft genome with n contigs $c_1, c_2, \ldots, c_n$ and $R_1, R_2, \ldots, R_k$ denote k reference genomes with weights $w_1, w_2, \ldots, w_k$, respectively. Contigs actually are fragmented linear DNA sequences with two *extremities*, called *head* and *tail*, respectively. Multi-CSAR performs the following steps to scaffold the contigs in the target genome T using the multiple reference genomes $R_1, R_2, \ldots, R_k$. First, Multi-CSAR utilizes CSAR to obtain a single reference-derived scaffold S_i of T based on each R_i, where $1 \le i \le k$. Second, Multi-CSAR constructs a *contig adjacency graph* $G = (V, E)$ such that there are two vertices c_j^h and c_j^t for representing the head and tail of each contig c_j, respectively, and there also is an edge for linking any two vertices if they are the extremities coming from the different contigs. An edge in E is said to be *supported* by a reference genome R_i if its two vertices are adjacent extremities from two distinct but continuous contigs in scaffold S_i. If an edge in E is supported by several reference genomes at the same time, then this edge receives a weight equal to the sum of the weights of all these supporting reference genomes. However, if an edge in E is not supported by any reference genome, then it has a weight of zero. Third, Multi-CSAR utilizes the Blossom V [20] to find a maximum weighted perfect matching M in G, where a subset of edges in G is called a *perfect matching* if every vertex in G is incident to exactly one edge in this subset. Let $C = \left\{ \left(c_j^t, c_j^h\right) | 1 \le j \le n \right\}$ and M' denote a subset of M (i.e., $M' \subseteq M$) with the minimum weight such that there is no cycle in $M' \cup C$. Finally, Multi-CSAR makes use of the edge connections in M' to scaffold the contigs of T. **Figure 1** displays an example for illustrating how the algorithm of Multi-CSAR works.

Note that CSAR was developed based on on a near-linear time algorithm [21] and Blossom V based on an $\mathcal{O}(n^4)$-time algorithm [20], where n is the number of vertices in a graph. Therefore, all the steps in the Multi-CSAR algorithm described previously can be implemented in polynomial time. In addition, Multi-CSAR utilizes the following *sequence identity-based weighting scheme* to automatically compute the weights $w_1, w_2, \ldots, w_k$ of the k reference genomes. First, Multi-CSAR applies either NUCmer or PROmer for identifying those *sequence markers* that actually are aligned regions between the target genome T and each reference genome R_i, where $1 \le i \le k$. Note that both NUCmer and PROmer come from the MUMmer package [22]. The main difference between NUCmer and PROmer is that the former finds the sequence markers directly on input DNA sequences, while the latter recognizes them on the six-frame protein translation of the input DNA sequences. Suppose that there are τ sequence markers, say $m_1, m_2, \ldots, m_\tau$, between T and R_i, and $L(m_j)$ and $I(m_j)$ are used to denote the alignment length of each m_j and its percent identity, respectively. Next, Multi-CSAR calculates the *weight* of each reference genome R_i by the formula $w_i = \sum_{j=1}^{\tau} L(m_j) \times I(m_j)$. The principle of the sequence identity-based weighting scheme is that the more similar the reference genome R_i is to the target genome T, the more weight R_i receives.

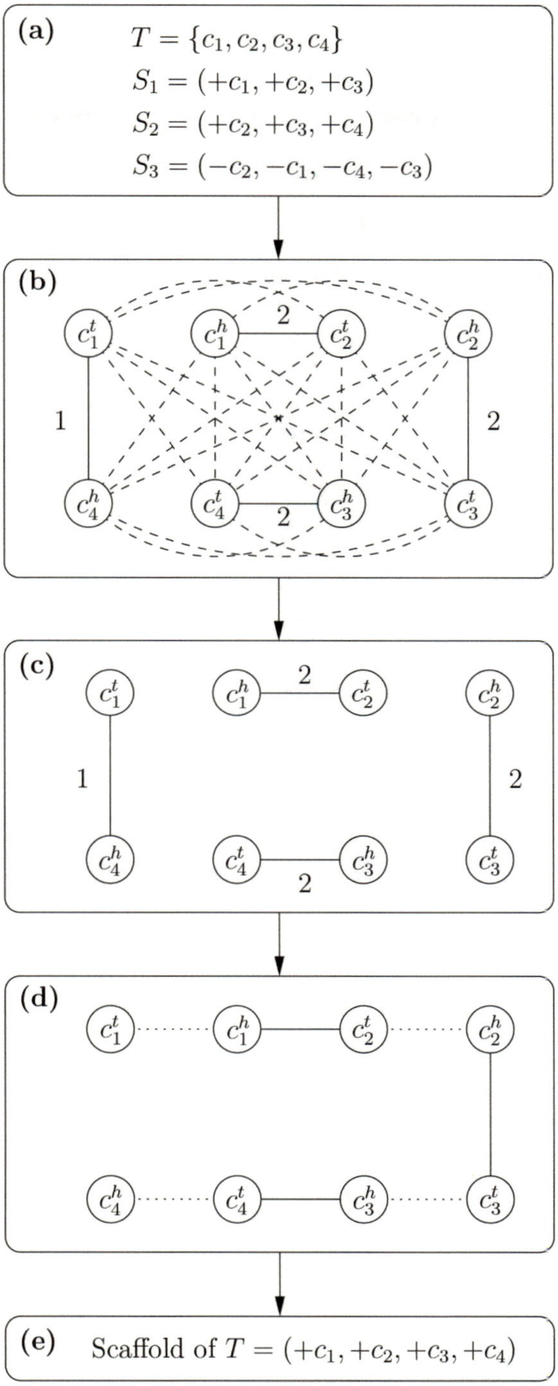

Figure 1.
Schematic workflow of multi-CSAR: (a) a target genome $T = \{c_1, c_2, c_3, c_4\}$ and three single reference-derived scaffolds $S_1 = (+c_1, +c_2, +c_3)$, $S_2 = (+c_2, +c_3, +c_4)$ and $S_3 = (-c_2, -c_1, -c_4, -c_3)$ that are supposed to be computed by applying CSAR on three reference genomes R_1, R_2 and R_3, respectively, with $w_1 = w_2 = w_3 = 1$. (b) The contig adjacency graph G constructed by using S_1, S_2 and S_3, where zero-weighted edges are denoted by dashed lines. (c) A perfect matching with maximum weight $M = \left\{(c_1^h, c_2^t), (c_2^h, c_3^t), (c_3^h, c_4^t), (c_4^h, c_1^t)\right\}$ derived by applying Blossom V on G. (d) $M' = \left\{(c_1^h, c_2^t), (c_2^h, c_3^t), (c_3^h, c_4^t)\right\}$ is obtained by removing edge (c_4^h, c_1^t) with minimum weight from M such that $M' \cup C$ contains no cycles, where the dotted lines denote the edges in C. (e) The final scaffold $(+c_1, +c_2, +c_3, +c_4)$ of T constructed based on the edge connections in M'.

3.2 Usage of multi-CSAR

Currently, Multi-CSAR offers a web server[1] with an easy-to-operate interface (see **Figure 2**) to the users. To run Multi-CSAR, the users first need to upload a target genome and one or more reference genomes in multi-FASTA format. If needed, the users can click the "plus" (respectively, "minus") button to add (respectively, remove) a reference genome field. Second, the users can determine whether or not to utilize the sequence identity-based weighting scheme provided by Multi-CSAR for automatically calculating the weights of reference genomes. If the weighting scheme is not used, then the weights of all the reference genomes are defaulted to one. Third, the users can choose either NUCmer or PROmer to identify sequence markers between the target genome and each of the reference genomes. Fourth, the users can enter an email address, which is optional, if they would like to run Multi-CSAR in a batch way. When running Multi-CSAR in this batch way, the users will be notified of the scaffolding result via email when the submitted job is finished by the web server of Multi-CSAR.

Multi-CSAR outputs its scaffolding results in four tab pages: (a) input data & parameters, (b) Circos plot validation, (c) dotplot validation, and (d) scaffolds of target. In the "Input data & parameters" page (see **Figure 3** for an example), Multi-CSAR simply shows the information of the input target and reference genomes, the user-specified program (either NUCmer or PROmer) for identifying their sequence markers, and whether the weighting scheme of reference genomes is used or not. By clicking on the links of the target and reference genomes in this page, Multi-CSAR

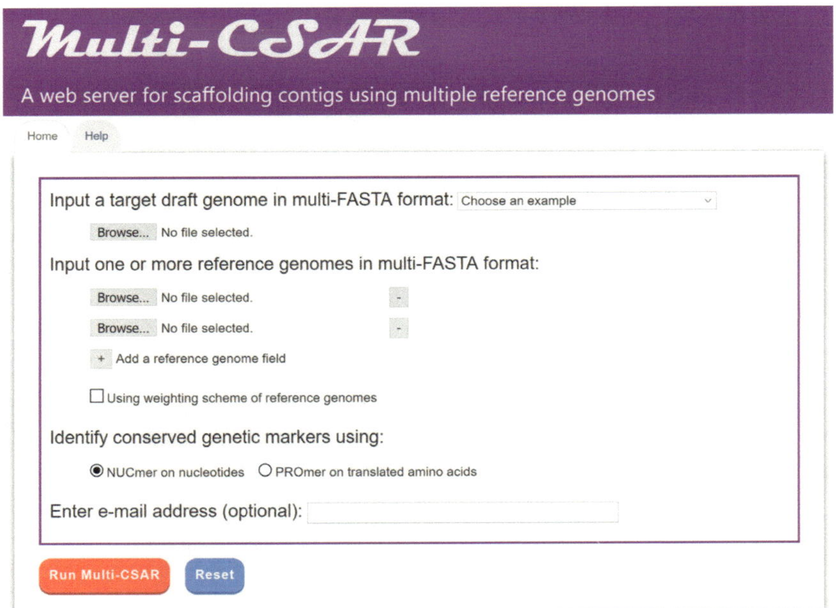

Figure 2.
Interface of multi-CSAR web server.

---

[1] The web server of Multi-CSAR is available at http://genome.cs.nthu.edu.tw/Multi-CSAR/.

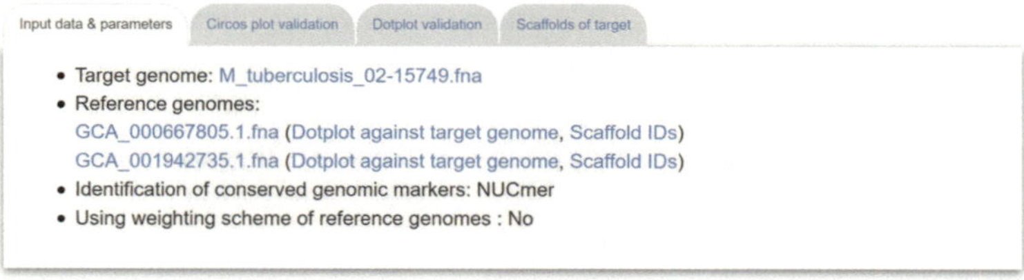

Figure 3.
A display of the "Input data & parameters" tab page.

will also display their input DNA sequences. By clicking on the link "Dotplot against target genome" on the reference genomes, Multi-CSAR will display a *dotplot* that allows the users to visually inspect sequence markers shared between un-scaffolded target genome and a reference genome. In the dotplot (see **Figure 4** for an instance), the un-scaffolded target genome and a selected reference genome are represented on the y and x axes, respectively. Note that the contigs and scaffolds in the dotplot are separated by horizontal and vertical dashed lines. Moreover, each forward (respectively, reverse) sequence marker is shown by a red (respectively, blue) line and its begin and end are represented by two unfilled circles. The users can sort the contigs of the input target genome based on their sizes by clicking on

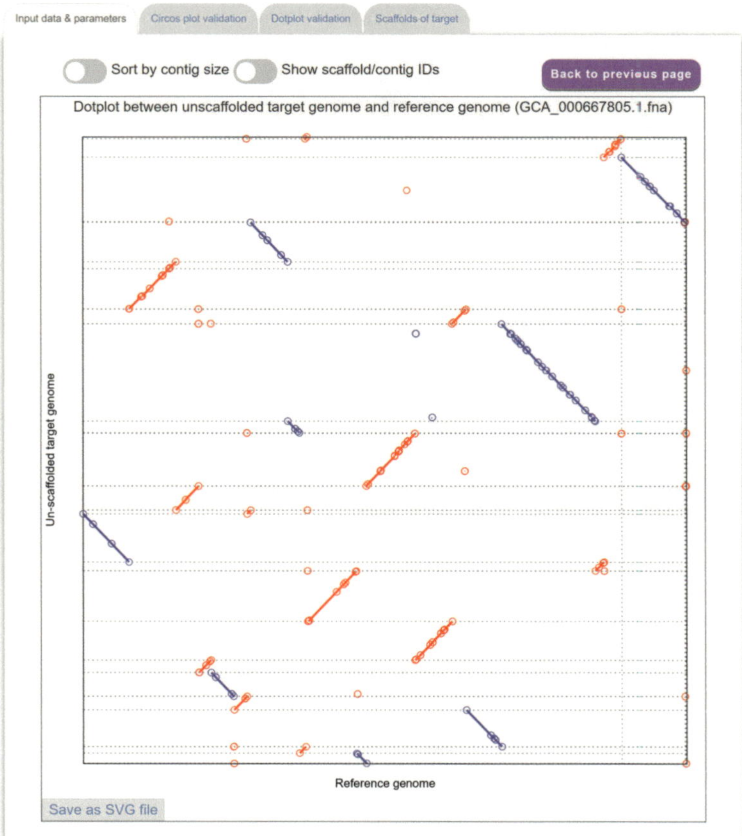

Figure 4.
A display of a dotplot between un-scaffolded target genome and a reference genome.

the toggle switch "Sort by contig size". The users also can show or hide the IDs of contigs and scaffolds used in Multi-CSAR by using the toggle switch "Show scaffold/contig IDs." The format of contig (respectively, scaffold) IDs begins with three-letter prefix CTG (respectively, SCF) followed by an underscore (_) and at least one digit (e.g., CTG_1 and SCF_1). In addition, the users can click the "Save as SVG file" button to download a copy of the dotplot in scalable vector graphics (SVG) format.

In the "Circos plot validation" page, (see **Figure 5** for an example), Multi-CSAR displays its total running time, as well as its scaffolding result by a Circos plot between scaffolded target genome and all reference genomes. In the initial Circos plot, the scaffolds of target genome (displayed in purple) and all the reference genomes (displayed in other colors) are arranged in a circle with the inner links connecting corresponding sequence markers between the target genome and each of reference genomes. The color of an inner link comes from the reference genome it connects. In the Circos plot, the number of crossing inner links can be viewed as a

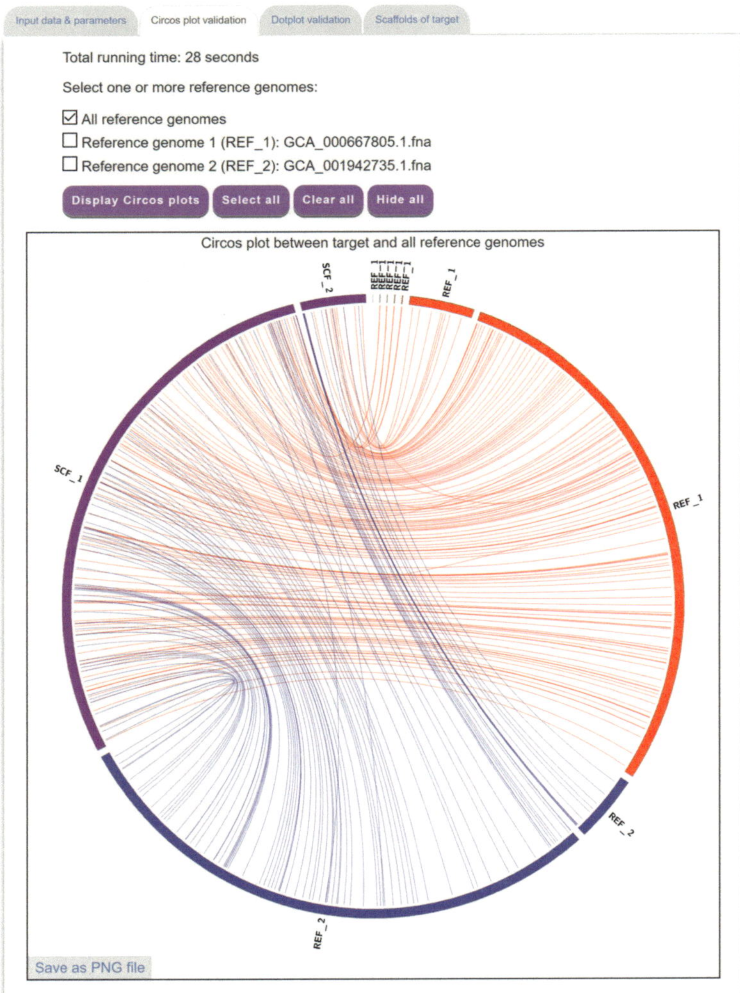

Figure 5.
A display of a Circos plot between scaffolded target genome and all reference genomes.

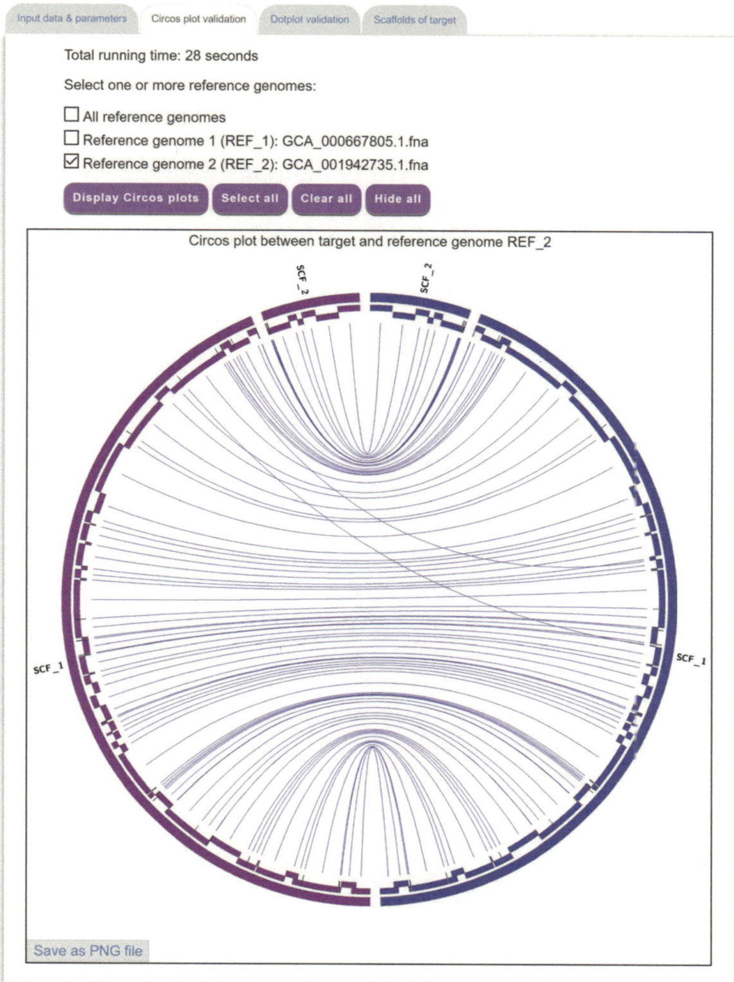

Figure 6.
A display of a Circos plot between scaffolded target genome and a selected reference genome, where the sequence markers are arranged in alternating layers along the two-layer inner circle.

accuracy measure for a scaffolding result. That is, if the contigs of the target genome are scaffolded well according to a reference genome, the number of crossing inner links between them should be low. For this purpose, Multi-CSAR allows the users to select any reference genome (by clicking the checkbox next to it) from the top of the tab page to display (by clicking the "Display Circos plot" button) its Circos plot against the scaffolded target genome (see **Figure 6** for an instance). In this Circos plot, the inner circle displays the sequence markers shared between the target genome and the selected reference genome. As demonstrated in **Figure 6**, the Circos plots of the scaffolding result are convenient and helpful for the users to visually validate whether the contigs of the target genome are properly scaffolded according to the reference genomes, as well as to visually identify whether there are any genome rearrangements between the scaffolded target and reference genomes. In addition, Multi-CSAR allows the users to the Circos plots of the scaffolds in portable network graphics (PNG) format by clicking the "Save as PNG file" button.

In the "Dotplot validation" page (see **Figure 7** for an example), Multi-CSAR displays its its scaffolding result by a dotplot between the scaffolded target genome and a selected reference genome (the default is the first reference genome). In fact, the matched sequence regions of sequence markers should be displayed from the bottom left to the top right in the dotplot (as shown in **Figure 7**) or from the top left to the bottom right, if the contigs from the target genome are scaffolded perfectly based on the selected reference genome. Showing the scaffolding result in the dotplot display is another way to conveniently help the users to visually verify whether the contigs of the target genome are scaffolded properly based on the reference genomes or not. The users can click the "Save as PNG file" button to download the dotplot of a scaffold in portable network graphics (PNG) format.

In the "Scaffolds of target" page (see **Figure 8** for an instance), Multi-CSAR displays its scaffolding result in tabular format for the purpose of allowing the users to view the scaffolds of the target genome in detail. The scaffolds in the table are sorted according to their sizes, which equals to the sum of contig sizes. In each scaffold, the ordered contigs, as well as their orientations (forward orientation denoted by 0 and reverse orientation by 1), sequences and lengths, are listed in a table. The users can click on the "Download scaffolds (.txt)" and "Download scaffolds (.csv)" buttons to download the scaffolds of the target genome in the tab-delimited text format and comma-delimited CSV format, respectively. In addition, the users can click on the "Download sequences" button to download the scaffold sequences in the text format, in which the sequences of contigs are separated by 100 Ns if they belong to the same scaffold.

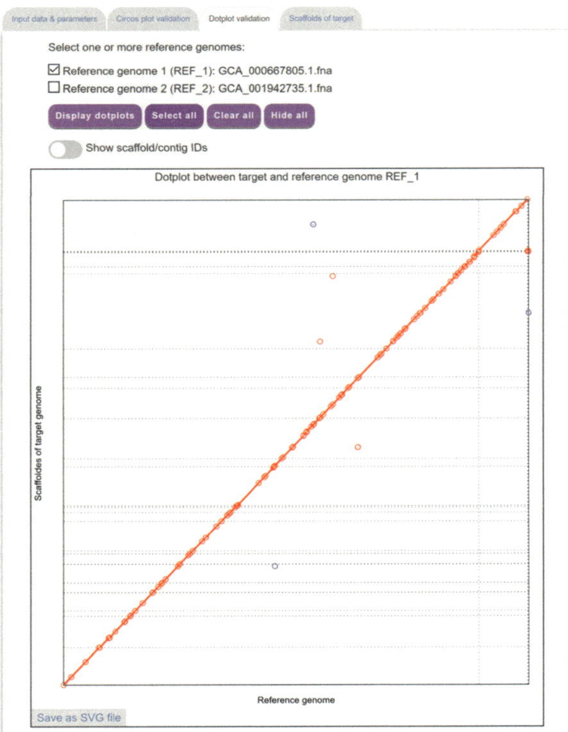

Figure 7.
A display of the "Dotplot validation" tab page.

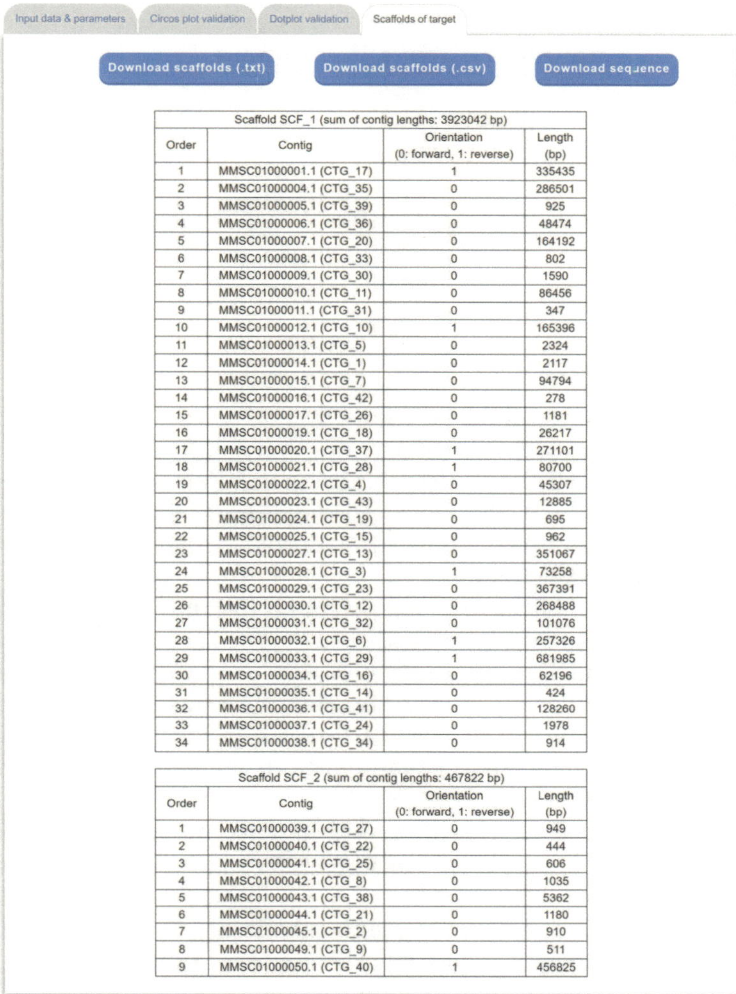

Scaffold SCF_1 (sum of contig lengths: 3923042 bp)			
Order	Contig	Orientation (0: forward, 1: reverse)	Length (bp)
1	MMSC01000001.1 (CTG_17)	1	335435
2	MMSC01000004.1 (CTG_35)	0	286501
3	MMSC01000005.1 (CTG_39)	0	925
4	MMSC01000006.1 (CTG_36)	0	48474
5	MMSC01000007.1 (CTG_20)	0	164192
6	MMSC01000008.1 (CTG_33)	0	802
7	MMSC01000009.1 (CTG_30)	0	1590
8	MMSC01000010.1 (CTG_11)	0	86456
9	MMSC01000011.1 (CTG_31)	0	347
10	MMSC01000012.1 (CTG_10)	1	165396
11	MMSC01000013.1 (CTG_5)	0	2324
12	MMSC01000014.1 (CTG_1)	0	2117
13	MMSC01000015.1 (CTG_7)	0	94794
14	MMSC01000016.1 (CTG_42)	0	278
15	MMSC01000017.1 (CTG_26)	0	1181
16	MMSC01000019.1 (CTG_18)	0	26217
17	MMSC01000020.1 (CTG_37)	1	271101
18	MMSC01000021.1 (CTG_28)	1	80700
19	MMSC01000022.1 (CTG_4)	0	45307
20	MMSC01000023.1 (CTG_43)	0	12885
21	MMSC01000024.1 (CTG_19)	0	695
22	MMSC01000025.1 (CTG_15)	0	962
23	MMSC01000027.1 (CTG_13)	0	351067
24	MMSC01000028.1 (CTG_3)	1	73258
25	MMSC01000029.1 (CTG_23)	0	367391
26	MMSC01000030.1 (CTG_12)	0	268488
27	MMSC01000031.1 (CTG_32)	0	101076
28	MMSC01000032.1 (CTG_6)	1	257326
29	MMSC01000033.1 (CTG_29)	1	681985
30	MMSC01000034.1 (CTG_16)	0	62196
31	MMSC01000035.1 (CTG_14)	0	424
32	MMSC01000036.1 (CTG_41)	0	128260
33	MMSC01000037.1 (CTG_24)	0	1978
34	MMSC01000038.1 (CTG_34)	0	914

Scaffold SCF_2 (sum of contig lengths: 467822 bp)			
Order	Contig	Orientation (0: forward, 1: reverse)	Length (bp)
1	MMSC01000039.1 (CTG_27)	0	949
2	MMSC01000040.1 (CTG_22)	0	444
3	MMSC01000041.1 (CTG_25)	0	606
4	MMSC01000042.1 (CTG_8)	0	1035
5	MMSC01000043.1 (CTG_38)	0	5362
6	MMSC01000044.1 (CTG_21)	0	1180
7	MMSC01000045.1 (CTG_2)	0	910
8	MMSC01000049.1 (CTG_9)	0	511
9	MMSC01000050.1 (CTG_40)	1	456825

Figure 8.
A display of the "Scaffolds of target" tab page.

4. Results and discussion

4.1 Testing datasets

The three multiple reference-based scaffolders Multi-CSAR, Ragout (version 1.0) and MeDuSa (version 1.6), we introduced in this chapter, were tested on a benchmark of five real bacterial datasets as shown in **Table 1**. In fact, these five testing datasets were originally prepared by Bosi et al. when they studied MeDuSa [16]. Basically, each testing dataset consists of a target draft genome to be scaffolded and two or more reference genomes that can be either complete or incomplete.

4.2 Evaluation metrics

For each testing dataset, Bosi et al. [16] also provided a *reference order* for the contigs of the target genome that can be used a truth standard to evaluate the

Organism	No. of replicons	No. of contigs	No. of references	Genome size (Mbp)	GC %
B. cenocepacia j2315	4	1223	4	8.05	65.9
E. coli K12	1	451	25	4.64	50.8
M. tuberculosis	1	116	13	4.41	65.6
R. sphaeroides 2.4.1	7	564	2	4.60	67.4
S. aureus	3	170	35	2.90	32.0

Table 1.
Summary of the five testing datasets.

multiple reference-based scaffolders. The evaluation metrics of the scaffolders include sensitivity, precision, *F-score*, genome coverage, NGA50, scaffold number and running time. Basically, sensitivity, precision and *F-score* are used to estimate the scaffold accuracy, genome coverage to estimate the scaffold coverage, and NGA50 and scaffold number to estimate the scaffold contiguity. Below, we introduced their detailed definitions.

Note that if any two contigs in a scaffold appear in continuous order and correct orientation in the reference order, then they are viewed as a *correct* join. Let S denote the result obtained by applying a scaffolder on a target genome T and P denote the number of all contig joins in the reference order. The number of the correct contig joins in S is then called as *true positive* (TP) and the number of the others (i.e., incorrect joins) as *false positive* (FP). In addition, the *sensitivity* of S is defined as TP/P, its *precision* as TP/(TP + FP), and its *F-score* as $(2 \times \text{sensitivity} \times \text{precision})/(\text{sensitivity} + \text{precision})$. Actually, *F-score* is a balanced measure between sensitivity and precision and *F-score* is high only when both sensitivity and precision are high.

Suppose that the target genome T contains only circular DNAs and C is a contig in S. If the both sides of C are joined correctly with two contigs, then the whole length of C will be counted in the genome coverage that will be defined later. If exactly one side of C is joined correctly with one contig, then half of the whole length of C will be counted. If the both sides of C are joined incorrectly with two contigs, then the whole length of C will be ignored. Based on the above discussion, the *genome coverage* of S is defined to be the ratio of the sum of the contig lengths counted according to the above-mentioned rules to the sum of all contig lengths. On the other hands, suppose that there are linear DNAs in the target genome T. Then in the reference order of each linear DNA, the first and last contigs have just one neighbor contig and thus only half of their lengths will be counted in the calculation of the genome coverage if these two contigs are correctly joined with neighbor contigs.

The NGA50 value of S is computed as follows [23]. First, the scaffolds of S are aligned with the complete sequence of the target genome T to find the mis-assembly breakpoints. Second, the scaffolds of S are broken at the mis-assembly breakpoints and their unaligned regions are also removed. Finally, the NGA50 value is equal to the NG50 value of the resulting scaffolds, which is the size of the shortest scaffold with longer and equal length scaffolds covering at least 50% of the target genome.

4.3 Comparison of multiple reference-based scaffolding results

All the three evaluated scaffolders Multi-CSAR, Ragout (version 1.0) and MeDuSa (version 1.6) were all run with their default parameters, except that a star

Scaffolder	Sen	Pre	*F*-score	Cov	NGA50	#Scaf	Time
Ragout	79.0	**92.5**	84.4	87.4	992,966	84	24.8
MeDuSa	78.2	81.9	80.0	83.3	671,001	26	3.8
Multi-CSAR (PROmer)	89.3	90.4	89.8	92.5	1,016,308	7	6.3
Multi-CSAR (NUCmer)	**89.6**	90.8	**90.2**	**93.2**	**1,038,257**	9	**1.7**

Table 2.
Average performance of the evaluated multiple reference-based scaffolders on the five testing datasets.

Scaffolder	Sen	Pre	*F*-score	Cov	NGA50	#Scaf	Time
Multi-CSAR (PROmer)	89.4	90.5	89.9	92.8	1,045,489	7	6.3
Multi-CSAR (NUCmer)	**89.9**	**91.3**	**90.6**	**93.5**	**1,046,288**	10	**1.7**

Table 3.
Average performance of multi-CSAR on the five testing datasets when using the sequence identity-based weighting scheme.

tree was used in Ragout to serve as the phylogenetic tree for each testing dataset because reliable phylogenetic trees were still unknown. **Table 2** displays their average performance results over the five bacterial datasets, by showing the values of sensitivity (Sen), precision (Pre), *F-score* and genome coverage (Cov) in percentage (%) and the size of NGA50 in base pairs (bp). In addition, **Table 2** shows the numbers of scaffolds computed by all evaluated scaffolders in the column '#Scaf' and their running times in minutes in the column 'Time'. The best result in each column of **Table 2** is shown in bold.

As shown in **Table 2**, Multi-CSAR running with NUCmer achieves the best sensitivity, *F-score*, genome coverage, NGA50 and running time, and the second best precision and scaffold number. On the other hands, Multi-CSAR running with PROmer has the best result in terms of scaffold number and the second best results in terms of sensitivity, *F-score*, genome coverage and NGA50. From the precision point of view, the performance of Ragout is the best among all the tested multiple reference-based scaffolders. However, the sensitivity of Ragout is substantially inferior to that of Multi-CSAR when either running with NUCmer or PROmer. This negative result also leads to that Ragout is much inferior to Multi-CSAR in the performance of *F-score*. Moreover, Ragout yields the worst results in terms of both scaffold number and running time. Compared Multi-CSAR and Ragout, MeDuSa gives the worst performance in sensitivity, precision, *F-score*, genome coverage and NGA50, although it has the second best performance in running time.

Table 3 shows the average performance results of Multi-CSAR on the five bacterial datasets when using the sequence identity-based weighting scheme, where the best performance in each column is also displayed in bold. As compared to the results of Multi-CSAR as shown in **Table 2**, several performance measures of Multi-CSAR can be further improved if it is run with the sequence identity-based weighting scheme of reference genomes, such as sensitivity, precision, *F-score*, genome coverage and NGA50.

5. Conclusions

Scaffolders are useful tools for sequencing projects to obtain more complete sequences of genomes being sequenced. In this chapter, we mainly introduced some state-of-the-art multiple reference-based scaffolders, such as Ragout, MeDuSa and

Multi-CSAR (improved extension of Multi-CAR), that can efficiently produce more accurate scaffolds of a target draft genome by referring to multiple complete and/or incomplete genomes of related organisms. By testing on five real prokaryotic datasets, Multi-CSAR outperforms Ragout and MeDuSa in terms of average sensitivity, precision, *F-score*, genome coverage, NGA50, scaffold number and running time. Currently, Multi-CSAR provides the users with a web interface that is intuitive and easy to operate. In addition, it displays its scaffolding result in a graphical mode that allows the users to visually validate the correctness of scaffolded contigs and in a tabular mode that allows the users to view the details of scaffolds.

Acknowledgements

This work was partially supported by Ministry of Science and Technology of Taiwan under grants MOST107-2221-E-007-066-MY2 and MOST109-2221-E-007-086.

Conflict of interest

The authors declare no conflict of interest.

Author details

Yi-Kung Shieh†, Shu-Cheng Liu† and Chin Lung Lu*
Department of Computer Science, National Tsing Hua University, Hsinchu, Taiwan

*Address all correspondence to: cllu@cs.nthu.edu.tw

† These authors are contributed equally.

References

[1] Goodwin S, McPherson JD, McCombie WR. Coming of age: Ten years of next-generation sequencing technologies. Nature Reviews. Genetics. 2016;**17**:333-351

[2] Pop M. Genome assembly reborn: Recent computational challenges. Briefings in Bioinformatics. 2009;**10**: 354-366

[3] Mardis E, McPherson J, Martienssen R, Wilson RK, McCombie WR. What is finished, and why does it matter. Genome Research. 2002;**12**:669-671

[4] Nagarajan N, Cook C, Di Bonaventura M, Ge H, Richards A, Bishop-Lilly KA, et al. Finishing genomes with limited resources: Lessons from an ensemble of microbial genomes. BMC Genomics. 2010;**11**:242

[5] van Hijum SA, Zomer AL, Kuipers OP, Kok J. Projector 2: Contig mapping for efficient gap-closure of prokaryotic genome sequence assemblies. Nucleic Acids Research. 2005;**33**:W560-W566

[6] Richter DC, Schuster SC, Huson DH. OSLay: Optimal syntenic layout of unfinished assemblies. Bioinformatics. 2007;**23**:1573-1579

[7] Assefa S, Keane TM, Otto TD, Newbold C, Berriman M. ABACAS: Algorithm-based automatic contiguation of assembled sequences. Bioinformatics. 2009;**25**:1968-1969

[8] Rissman AI, Mau B, Biehl BS, Darling AE, Glasner JD, Perna NT. Reordering contigs of draft genomes using the mauve aligner. Bioinformatics. 2009;**25**:2071-2073

[9] Husemann P, Stoye J. r2cat: Synteny plots and comparative assembly. Bioinformatics. 2010;**26**:570-571

[10] Galardini M, Biondi EG, Bazzicalupo M, Mengoni A. CONTIGuator: A bacterial genomes finishing tool for structural insights on draft genomes. Source Code for Biology and Medicine. 2011;**6**:11

[11] Munoz A, Zheng C, Zhu Q, Albert VA, Rounsley S, Sankoff D. Scaffold filling, contig fusion and comparative gene order inference. BMC Bioinformatics. 2010;**11**:304

[12] Dias Z, Dias U, Setubal JC. SIS: A program to generate draft genome sequence scaffolds for prokaryotes. BMC Bioinformatics. 2012;**13**:96

[13] Lu CL, Chen KT, Huang SY, Chiu HT. CAR: Contig assembly of prokaryotic draft genomes using rearrangements. BMC Bioinformatics. 2014;**15**:381

[14] Chen KT, Liu CL, Huang SH, Shen HT, Shieh YK, Chiu HT, et al. CSAR: A contig scaffolding tool using algebraic rearrangements. Bioinformatics. 2018;**34**:109-111

[15] Kolmogorov M, Raney B, Paten B, Pham S. Ragout: A reference-assisted assembly tool for bacterial genomes. Bioinformatics. 2014;**30**:i302-i309

[16] Bosi E, Donati B, Galardini M, Brunetti S, Sagot MF, Lio P, et al. MeDuSa: A multi-draft based scaffolder. Bioinformatics. 2015;**31**:2443-2451

[17] Chen KT, Chen CJ, Shen HT, Liu CL, Huang SH, Lu CL. Multi-CAR: A tool of contig scaffolding using multiple references. BMC Bioinformatics. 2016; **17**:469

[18] Chen KT, Shen HT, Lu CL. Multi-CSAR: A multiple reference-based contig scaffolder using algebraic rearrangements. BMC Systems Biology. 2018;**12**:139

[19] Pagani I, Liolios K, Jansson J, Chen IMA, Smirnova T, Nosrat B, et al. The genomes OnLine database (GOLD) v.4: Status of genomic and metagenomic projects and their associated metadata. Nucleic Acids Research. 2012;**40**:D571-D579

[20] Kolmogorov V. Blossom V: A new implementation of a minimum cost perfect matching algorithm. Mathematical Programming Computation. 2009;**1**:43-67

[21] Lu CL. An efficient algorithm for the contig ordering problem under algebraic rearrangement distance. Journal of Computational Biology. 2015; **22**:975-987

[22] Kurtz S, Phillippy A, Delcher AL, Smoot M, Shumway M, Antonescu C, et al. Versatile and open software for comparing large genomes. Genome Biology. 2004;**5**. Available from: https://genomebiology.biomedcentral.com/articles/10.1186/gb-2004-5-2-r12

[23] Gurevich A, Saveliev V, Vyahhi N, Tesler G. QUAST: Quality assessment tool for genome assemblies. Bioinformatics. 2013;**29**:1072-1075

Section 4

Computational Biology and Chemical Monitoring

Chapter 5

Biological Evaluation and Molecular Docking Studies of Benzalkonium Ibuprofenate

Kodakkat Parambil Safna Hussan,
Mohamed Shahin Thayyil, Thaikadan Shameera Ahamed
and Karuvanthodi Muraleedharan

Abstract

The third-generation ionic liquids (ILs), which are being used to produce double active pharmaceutical ingredients (d-APIs) with tunable biological activity along with novel performance, enhancement, and delivery options, have been revolutionizing the area of drug discovery since the past few decades. Herein we report the in vitro antibacterial and anti-inflammatory activity of benzalkonium ibuprofenate (BaIb) that are being used as in-house d-API, with a particular focus on its interaction with respective protein target through molecular docking study. The evaluation of the biological activity of BaIb with the antibacterial and anti-inflammatory target at the molecular level revealed that the synthesized BaIb could be designed as a potential double active drug since it retains the antibacterial and anti-inflammatory activity of its parent drugs, benzalkonium chloride (BaCl) and sodium ibuprofenate (NaIb), respectively.

Keywords: benzalkonium ibuprofenate, double active pharmaceutical ingredient, molecular docking studies, anti-inflammatory, antibacterial activities

1. Introduction

In the pharmaceutical field, ionic liquids (ILs) by salification of drugs were widely applied to improve the performance of drugs on its oral administration, especially, their solubility, bioavailability, and stability. The research on the antimicrobial activity of ILs is a growing field because of its unprecedented flexibility for chemical diversity in a severely drained arsenal of antimicrobial. The third-generation ionic liquids give us the freedom to tune the biological properties in addition to its physical and chemical properties. The proper selection of ions with synergetic effects may result in the formation of double active pharmaceutical ingredient (d-API). Thus the d-APIs are composed of asymmetric organic ions, which prevent the formation of the stable crystal lattice and are liquid at unusually low temperatures. Such d-APIs can be used for the ailment situations where the two activities are required. This strategy will reduce the excess in taking of unwanted chemicals and will enhance the solubility and bioavailability [1, 2].

Benzalkonium ibuprofenate (BaIb) is a double active pharmaceutical ingredient designed by combining benzalkonium cations with ibuprofen. Benzalkonium chloride

(BaCl) is a potential antibacterial drug and sodium ibuprofen is a prospective anti-inflammatory drug [3]. It is important to evaluate the pharmaceutical profiles of the d-API to confirm the retainity of the biological activities of the parent drugs. We have synthesized a d-API, benzalkonium ibuprofenate, carried out its quantum mechanical calculations using density functional theory, characterized different experimental techniques, and reported glass-forming ability in earlier works [4, 5]. Now, in this work, the biological evaluations, in particular, the antibacterial and anti-inflammatory activities, were performed and the results with the activity of parent drugs, BaCl and sodium ibuprofenate (NaIb), compared. The molecular docking of all the samples was done to get a better understanding of the mode of interaction between the drugs and respective targeted proteins and to trace their binding pores and cites.

2. Materials and method

2.1 Materials

The benzalkonium chloride and sodium ibuprofenate were purchased from Sigma-Aldrich (USA). Cell culture plastic flasks, test tubes, and culture plates were purchased from Borosil (India).

2.2 Experimental

2.2.1 Synthesis

The double active pharmaceutical ingredient, benzalkonium ibuprofenate, was synthesized using stoichiometric metathesis reaction. Solid (1 mmol) BaCl and NaIb were dissolved in 50 mL distilled water, each taken in two beakers and stirred separately with gentle heating (40–60°C). Then the two solutions were mixed together and again stirred for another 30 min with heating (around 80°C) and then cooled to room temperature. 60 mL of chloroform was added to separate the organic and inorganic part; then the chloroform phase was washed with cold distilled water until it removes the inorganic salt completely. $AgNO_3$ test was used to confirm the absence of chloride anions in the product. This is followed by continuous washing of the chloroform phase with deionized (DI) water until the water washings tested negative for NaCl or NaBr via $AgNO_3$ test. The chloroform was then evaporated using rotary evaporator, and the BaIb was dried under high vacuum for 12 h with gentle heating (50–60°C) [4, 6].

The double active pharmaceutical ingredient, BaIb was characterized well using Nuclear Magnetic Resonance using Bruker Avance III, 400MHzwith a 9.4 Tesla super-conducting magnet in an operating temperature at 309 K, Fourier transform infrared spectroscopy JASCO FTIR-4100 spectrophotometer, Fourier Transform-Raman spectroscopy and UV–visible spectroscopy using Jasco UV–Visible Spectrophotometer model V-550 (USA) and are reported in earlier works [4, 5].

2.2.2 Biological evaluation

2.2.2.1 Antibacterial activity

The synthesized double active pharmaceutical ingredient BaIb was screened against Gram-negative bacteria to confirm the retainity of the biological activity of its parent drug BaCl, which is a potential germicide. For this study, we have chosen *Pseudomonas aeruginosa* and *E. coli* as Gram-positive, while DMSO is considered as

Gram-negative. Paper disk method was used for the in vitro analysis. The experiment was started with the preparation of nutrient agar (28 g in 100 mL) and was sterilized by autoclaving. This nutrient agar media was cooled without solidifying and incubated overnight. *P. aeruginosa* and *E. coli* were prepared. 15–20 mL of this media was poured into sterilized Petri plate and allowed to solidify. Then sterilized disks were placed in this solidified agar plate; each plate contains three disks. 10 μL negative controls on to one disk, 10 μL standard parent drugs on to the second disk, and 10 μL samples on to the third disk were added and appropriately labeled. After incubating the Petri dishes in 310 K for 24 h, the plates were checked for the zone of inhibition, and the zone diameter was measured using a scale [4, 7].

$$Formula\ for\ ABA = \frac{sample\ diameter}{+ve\ control} \times 100 \quad (1)$$

2.2.2.2 Anti-inflammatory activity

2.2.2.2.1 In chick albumin

The anti-inflammatory activity of BaIb and NaIb was done using an egg albumin. For this, a reaction mixture of 5 mL was made with fresh hen's egg (0.2 mL) and phosphate buffer saline with pH = 6.4 of 2 mL with varying concentrations of extract for preparing the concentrations of 100, 200, 300, 400, and 500 μg/mL. The same steps were repeated for the preparation of double distilled water, which served as control. Then the prepared mixtures were incubated in BOD incubator (Labline Technologies, India) at 210 ± 2 K for 15 min; after that, it is heated to 343 K and was hold for 5 min. The absorbance was measured after incubating using Shimadzu (Japan), UV 1800 at 660 nm. At the final concentration of 100, 200, 300, 400, and 500 μg/mL, acetyl salicylic acid was used as reference drug. The protein denaturation inhibition percentage was calculated using the following formulae:

$$\%inhibition = \frac{Abs_{control} - Abs_{test}}{Abs_{control}} \times 100 \quad (2)$$

2.2.2.2.2 In human serum albumin

In addition, the synthesized drug BaIb was screened for its anti-inflammatory activity, and its efficiency with the parent drug NaIb and drug diclofenac was compared. For this, blood samples from a healthy donor (male) were collected and mixed with sterilized Alsever's solution before centrifuging it at 3000 rpm. The suspension of packed cells was made with isoline. This suspension with phosphate buffer, hyposaline, was mixed with diclofenac at varying concentration, where the distilled water is taken as control while diclofenac as standard. Then the mixtures were incubated for 30 min at 303 K and centrifuged. Spectrophotometric analysis was used for hemolysis at 560 nm and its percentage recorded [4, 8, 9].

2.3 Computational

The input structures of BaCl (PubChem: 2330), NaIb (PubChem: 5338317), and BaIb (PubChem: 86612072) were taken from the PubChem database [9] and optimized using density functional theory with B3LYP level of theory and 631-G+(d,p) [10] basis sets using Gaussian software packages [11]. Further, the molecular docking was also conducted using Schrodinger Maestro software package [10]. The optimized structures and downloaded Protein Data Bank (PDB) files [12] of proteins were used for molecular docking studies.

The Schrodinger's Glide module was used for docking analysis of the present work. Glide offers the full range of speed vs. accuracy options, from the HTVS (high-throughput virtual screening) mode for efficiently enriching million compound libraries, to the SP (standard precision) mode for reliably docking tens to hundreds of thousands of ligand with high accuracy, and to the extra precision (XP) mode where further elimination of false positives is accomplished by more extensive sampling and advanced scoring, resulting in even higher enrichment. Many researchers carried out extensive comparisons of several docking programs and scoring functions using an extensive data set of pharmaceutically attractive targets and active compounds [13–18]. All the study leads to the same result that Glide XP methodology was shown to yield enrichments superior to the alternative methods consistently. Glide SP scoring also shows improvement as compared to the scoring in GOLD and DOCK. The drawbacks of Glide come from the fact that it's increasing computational time. From computational efficiency, the CPU time required on average for Glide XP calculations (7.0 min per ligand) is larger than other methods except for the most accurate version of Goldscore (8.5 min per ligand). This extra cost for Glide XP is the trade-off for the higher enrichment factors obtained. Glide SP delivers the second best overall enrichment performance while providing a considerable speedup (0.42 min per ligand) as compared to all approaches except for the fast version of GOLD Chemscore setting.

2.3.1 Molecular docking studies

The structure-based drug design always promotes the in silico method for molecular docking before going to lab screening. In silico methods can site the binding pores and predict the mechanism of protein-ligand interactions as well as target binding.

Moreover, the analysis and interpretation of the binding behavior play a crucial role in rational drug designs and in elucidating fundamentals of biochemical processes. The antibacterial activity of BaCl and BaIb was studied using LpxC enzymes since the enzyme LpxC places an important role in the lipid A biosynthesis. Lipid A acts as a hydrophobic membrane of lipopolysaccharide (LPS) in the outer leaflet of the outer membrane of Gram-negative bacteria; however, the bacteria is with a defective lipid. A synthesis reduces its hydrophobicity and shows increased membrane permeability, which in turn increase the sensitivity to the antibiotics, and hence, it results in cell death. For this work, we have selected LpxC from *Escherichia coli* and *P. aeruginosa*. In the same way, the targeted protein for the anti-inflammatory activity of NaIb and BaIb were studied using human serum albumin (HSA).

Thus the structures of proteins used in this work were downloaded from the Protein Data Bank [12]. The detailed information of the selected proteins, their PDB IDs, inbuilt inhibitor, X-ray resolution, etc., were given in **Table 1**. Molecular docking study has been carried out by the Glide docking program [19–21] provided by Schrodinger suite. Protein preparation is done by using the Protein Preparation Wizard module of Glide [22, 23]. Initially, all the protein structures must be preprocessed to be used as a receptor for docking. Some of the typical operations in preprocessing include (i) addition of hydrogen atoms, (ii) assignment of atomic charges, and (iii) elimination of water molecules that are not involved in ligand binding. Missing chains and loops can also be added if necessary. Preprocessed protein was optimized with PROPKA and then minimized with OPSL3 force field function, which is followed by a convergence of heavy atoms of RMSD 0.3 Å.

Then, the Glide's receptor grid generation wizard was used to generate a three-dimensional (3D) grid with a maximal size of 20 × 20 × 20 Å with 0.5 Å spacing.

Sl. no.	Protein	Organism	PDB-ID	Inbuilt inhibitor	X-ray resolution	Ligands	Activity
1	LpxC	*Escherichia coli*	3P3G	3p3	1.65 Å	BaCl, BaIb	Antibacterial
2	LpxC	*Pseudomonas aeruginosa*	5U3B	NVS	2.00 Å	BaCl, BaIb	Antibacterial
3	HSA	*Homo sapiens*	2BXG	Ibuprofen	2.70 Å	NaIb, BaIb	Anti-inflammatory

Table 1.
Detailed information regarding the proteins under study.

There is enough option to apply any constraints such as precision constraints, H-bond constraint, etc., in the receptor grid generation wizard. At last, flexible docking was performed with extra precision docking mode in Glide docking module.

3. Results and discussions

3.1 Antibacterial activities of BaCl and BaIb

3.1.1 In vitro studies

As we know, benzalkonium chloride is a prominent germicide widely used in medicinal chemistry [24]. It is mandatory to confirm the retainity of BaCl's antibacterial effect in the synthesized d-API, BaIb. The inhibition zone method using agar diffusion was used to screen the antibacterial activity of the prepared as well as parent drug against *P. aeruginosa* and *E. coli* bacterial strains [25]. The percentage of inhibition and its diameter are listed in **Table 2.** The results emphasized that BaIb retains the antibacterial activity of parent drug BaCl, though it had less inhibitory action against *E. coli* and *P. aeruginosa* than the parent drug [25].

3.1.2 Molecular docking studies with LpxC (E. coli)

Here, molecular docking of the parent and daughter drugs, BaCl and BaIb with *Escherichia coli* LpxC/LPC-009 complex, has been employed to trace its binding pore and binding affinity. The docking scores and binding free energies of lowest energy pose of its inbuilt inhibitor, 3p3 and the samples under study, BaCl, and BaIb in active sites on chain A of the LpxC/LPC-009 X-ray crystal structures have been computed after deleting the unwanted ligands and amino acids (So4 at 501, 502, 504, 505, 506; dimethyl sulfide (DMS) at 701; and UKW) using Schrodinger Glide module and are given in **Table 3.** The docking result points out that the drugs BaCl and BaIb show considerable binding affinity scores compared to the inbuilt ligand. However, interestingly the d-API BaIb exhibits high docking score compared to the parent drug BaCl, which emphasize that the interaction between the ligand and protein increases on double active formation with ibuprofen.

Figure 1 demonstrates the three-dimensional protein-ligand interaction of the three samples under study in the dynamic site of LpxC/LPC-009 obtained from graphical interface Maestro. All the ligands are found to be buried in the deep binding pocket of LpxC/LPC-009 in the same way. The d-API BaIb interacts with the active site's amino acids of the protein by H-bonding, which is depicted in red and dotted lines.

The diameter of zone of inhibition (mm)			Percentage of inhibition (%)		
	E. coli	*P. aeruginosa*		*E. coli*	*P. aeruginosa*
Standard BaCl	19	38	Standard BaCl	21.11	42.22
Sample BaIb	14	31	Sample BaIb	15.55	34.44
Negative (DMSO)	0	0	Negative (DMSO)	0	0

Table 2.
Preliminary in vitro antibacterial screening activity of BaIb.

Compound	Schrodinger software	
	Glide docking score (kcal/mol)	Glide ligand efficiency
3P3	−13.682	−0.507
BaCl	−3.234	−0.147
BaIb	−6.315	−0.175

Table 3.
Docking scores and binding free energies of inbuilt inhibitor 3P3, BaCl, and BaIb to the LpxC/LPC-009 using Schrodinger Maestro software.

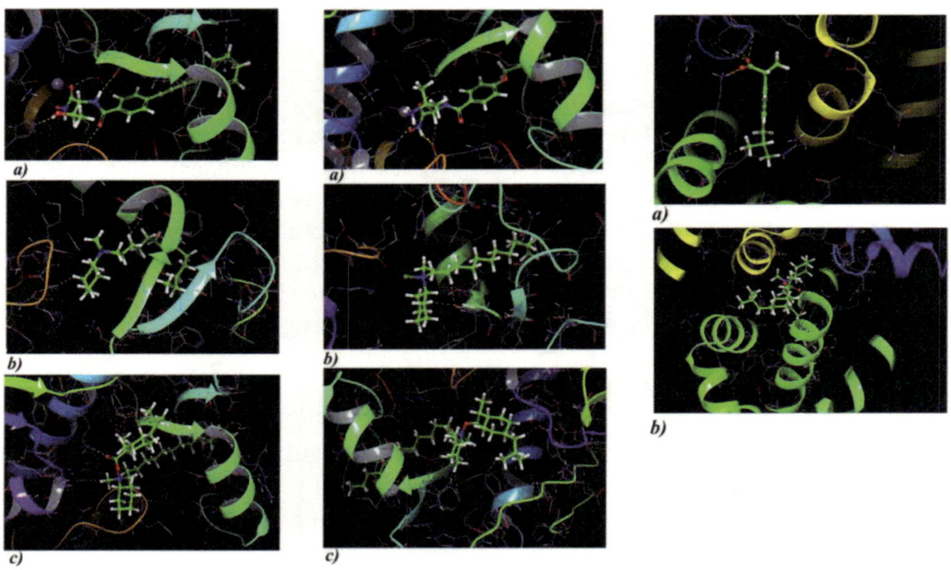

Figure 1.
Three-dimensional (3D) protein-ligand interactions diagram using Schrodinger software (I) with LpxC protein of E. coli *using (a) inbuilt ligand 3P3, (b) parent ligand BaCl, and (c) double active pharmaceutical ingredient BaIb; (II) with LpxC* (P. aeruginosa) *using (a) inbuilt ligand 3P3, (b) parent ligand BaCl, and (c) double active pharmaceutical ingredient BaIb; and (III) with human serum albumin using (a) inbuilt ligand ibuprofen and (b) double active pharmaceutical ingredient BaIb.*

In addition to the 3D binding orientations of the ligands in the protein, the docking results provide further insights into selective interactions of the ligands with the *E. coli* LpxC in the 2D image, as shown in **Figure 2**. The ligands were encompassed by active site amino acids THR191, PHE192, SER211, PHE212, CYS214, LYS239, HIS238, HIS265, etc., of LpxC. The co-crystallized ligand 3P3 occupied the deep cavity by forming three hydrogen bonds and one π-π interaction with active site amino acid PHE212. Though the parent ligand BaCl occupied the deep cavity of

LpxC with the support of salt bridges between them, the daughter ligand BaIb formed one hydrogen bond with the active amino acid LYS239.

Despite the binding energy and docking score difference between parent and daughter drugs, BaCl and BaIb were well occupied in the binding site of the LpxC protein as similar to the native ligand 3P3 by forming a hydrogen bond with active site amino acids. Besides, the binding energy and docking score emphasize that the d-API has a higher binding affinity with LpxC than the parent drug BaCl. This indicates that BaIb can be considered as a potential inhibitor of LpxC protein with antibacterial activity.

3.1.3 Molecular docking studies with LpxC (P. aeruginosa)

Here, molecular docking of the parent and daughter drugs, BaCl and BaIb with LpxC protein of *Pseudomonas aeruginosa*, has been employed to trace the nature of its binding interaction and binding affinity value. The docking scores and binding free energies of lowest energy pose of its inbuilt inhibitor, NVS-LpxC-01 and the samples under study, BaCl, and BaIb in active sites on chain B of the NVS-LpxC X-ray crystal structures, have been computed after deleting the unwanted ligands and amino acids and including the Zn^{2+} using Schrodinger Maestro software and are given in **Table 4**. The docking result points out that the drugs BaCl and BaIb show considerable binding affinity scores compared to the inbuilt ligand. However,

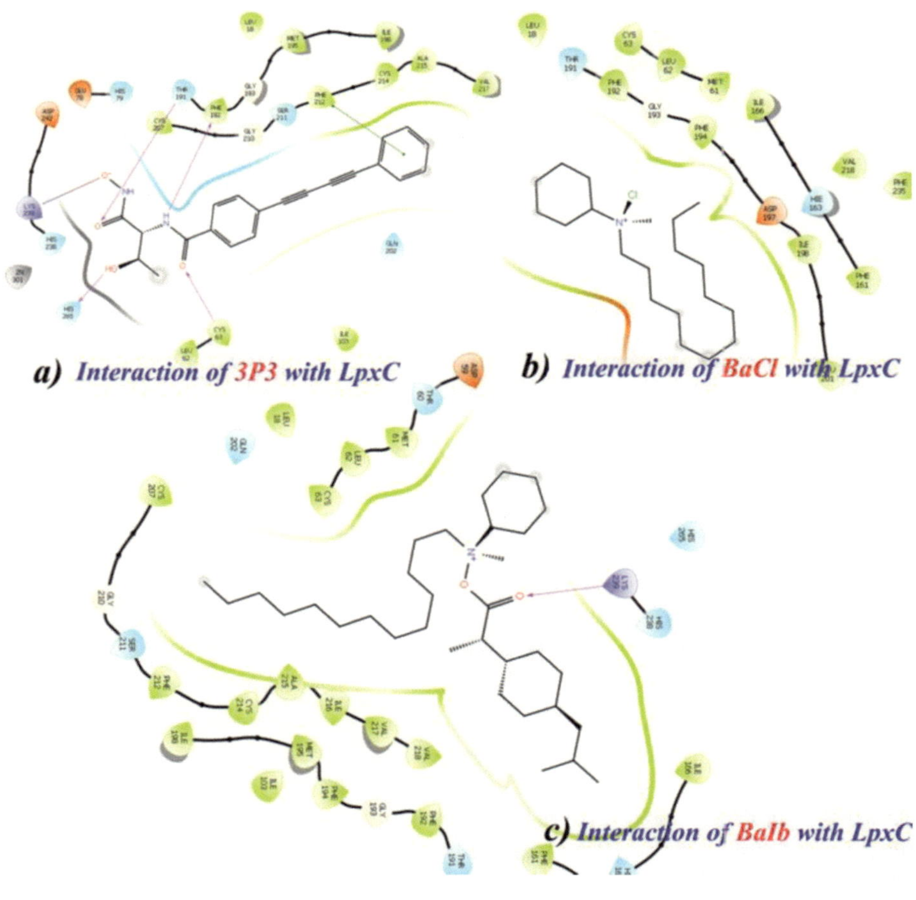

Figure 2.
Schematic representations of ligand-protein interaction and binding interaction using stick mode. (a) 3P3 with LpxC, (b) BaCl with LpxC, and (c) BaIb with LpxC.

Compound	Schrodinger software	
	Glide docking score (kcal/mol)	**Glide ligand efficiency**
NVS	−11.095	−0.482
BaCl	−4.540	−0.206
Balb	−5.494	−0.153

Table 4.
Docking scores and binding free energies of inbuilt inhibitor NVS, BaCl, and Balb to the LpxC/LPC-009 using Schrodinger Maestro software.

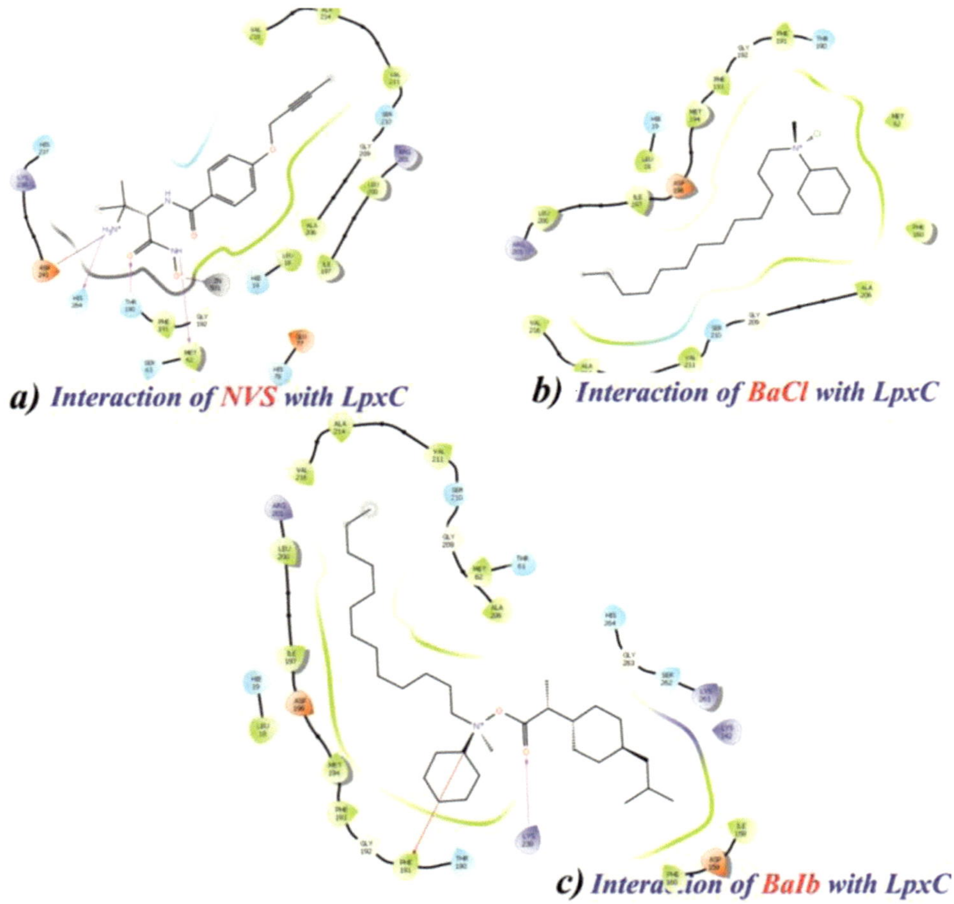

Figure 3.
Schematic representations of ligand-protein interaction and binding interaction using stick mode. (a) 3P3 with LpxC, (b) BaCl with LpxC, and (c) Balb with LpxC using Schrodinger software.

interestingly the d-API Balb exhibits high docking score compared to the parent drug BaCl, which emphasizes that the interaction between the ligand and protein increased on double active formation with ibuprofen.

Figure 1 demonstrates the three-dimensional protein-ligand interaction of three samples under study in the active site of LpxC in complex obtained from graphical interface Maestro. All the ligands are found to be buried in the deep binding pocket of LpxC in the complex in an indistinguishable way. The d-API Balb interact with the active site's amino acids of the protein by H-bonding, which is depicted in red and dotted lines.

In addition to the 3D binding orientations of the ligands in the protein, the docking results provide further insights into selective interactions of the ligands with the *Pseudomonas aeruginosa* LpxC in the 2D image as shown in **Figure 3**. The ligands were surrounded by active site amino acids THR190, GLY192, PHE191, SER211, PHE193, MET194, ASP196, DLE197, LEU200, ARG201, VAL216, etc., of LpxC. The co-crystallized ligand NVS occupied the deep cavity by forming five hydrogen bonds in addition to its salt bridge. Though the parent ligand BaCl occupied the deep cavity of LpxC with the support of salt bridges between them, the daughter ligand BaIb formed one hydrogen bond with the active amino acid LYS238 and a π-cation interaction with active site amino acid PHE191.

Despite the binding energy and docking score difference of parent as well as the daughter drugs, BaCl and BaIb were well occupied in the binding site of the LpxC protein as similar to the native ligand NVS with hydrogen bonding with active site amino acids. Also, the binding energy and docking score emphasize that the d-API has a higher binding affinity with LpxC than the parent drug BaCl. This indicates the BaIb can be considered as a potential inhibitor of LpxC protein with antibacterial activity.

3.2 Anti-inflammatory activities of NaIb and BaIb

3.2.1 In vitro studies using chick albumin

The anti-inflammatory properties of the synthesized drug BaIb and parent drug NaIb were studied in chick albumin membrane. **Figure 4** depicts the bar diagram of the percentage inhibition of inflammation against the concentration of NaIb and BaIb. Sample NaIb and BaIb have almost the same inhibitory activity in all concentration, which is around 95%. Thus from this analysis, one can confirm that BaIb retains the anti-inflammatory activity of NaIb and states.

3.2.2 In vitro studies using human serum albumin

The anti-inflammatory properties of the synthesized drug BaIb, parent drug NaIb, and drug diclofenac as standard samples were given in **Table 5**. Among samples provided, diclofenac shows maximum inhibitory activity, whereas the NaIb and BaIb are less active than the reference compound, but still, its activity is significant as an inflammatory agent. This fact suggests that the anti-inflammatory activity of NaIb was retained in the double active pharmaceutical ingredient. Thus the in vitro study confirms that the synthesized BaIb is a double active pharmaceutical ingredient by retaining the biological activities of the parent drugs.

3.2.3 Molecular docking studies of NaIb and BaIb with HSA

Here, molecular docking of the parent and daughter drugs, NaIb and BaIb, respectively, with human serum albumin complex has been employed to analyze its binding mode and binding affinity value. The docking scores and binding free energies of lowest energy pose of its inbuilt inhibitor, ibuprofen and the BaIb in active sites on chain A of the 2BXG X-ray crystal structures, have been computed after deleting the unwanted ligands and amino acids using Schrodinger Maestro software and are given in **Table 6**. The docking result points out that the drugs ibuprofen and BaIb show considerable binding affinity scores compared to the inbuilt ligand. However, interestingly the d-API BaIb exhibited almost similar docking score to the parent drug Ib, which emphasizes that the interaction between the ligand and protein is still functional on double active formation with ibuprofen.

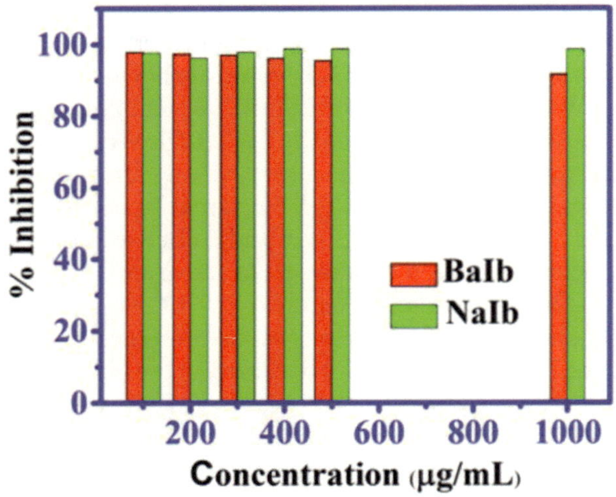

Figure 4.
Plot of percentage inhibition of inflammation against the concentration of Nalb and Balb studied in chick albumin membrane.

Percentage of inhibition of hemolysis (%)	
Diclofenac	92.16
Standard sample Nalb	88.47
Sample Balb	88.59

Table 5.
Preliminary in vitro anti-inflammatory properties of Balb.

Compound	Schrodinger software	
	Glide docking score (kcal/mol)	Glide ligand efficiency
Ibuprofen	−5.435	−0.362
Balb	−3.554	−0.099

Table 6.
Docking scores and binding free energies of inbuilt inhibitor ibuprofen and Balb to the 2BXG using Schrodinger Maestro software.

Figure 1 demonstrates the three-dimensional protein-ligand interaction of three samples under study in the dynamic site of 2BXG obtained from graphical interface Maestro. All the ligands are found to be well occupied in the deep binding pocket of 2BXG in the same way. The d-API Balb interact with the active site's amino acids of the protein by H-bonding, which is depicted in red and dotted lines. Interaction of amino acids at the active site of HAS with the studied compound is displayed in the 2D image (**Figure 5**). The ligands were encircled by amino acids like SER480, LBU481, VAL482, ASN483, PHE205, ARG209, ALA210, ALA213, etc., of HSA. The co-crystallized ligand ibuprofen occupies the deep cavity by forming three hydrogen bonds with active sites of amino acids SER480, LBU481, VAL482, and one π-anion interaction with active site amino acid LYS351. The daughter ligand Balb forms only one hydrogen bond with the amino acid LYS239.

Despite the binding energy and docking score difference of parent as well as the daughter drug, the daughter drug, Balb, was well occupied in the binding site of the

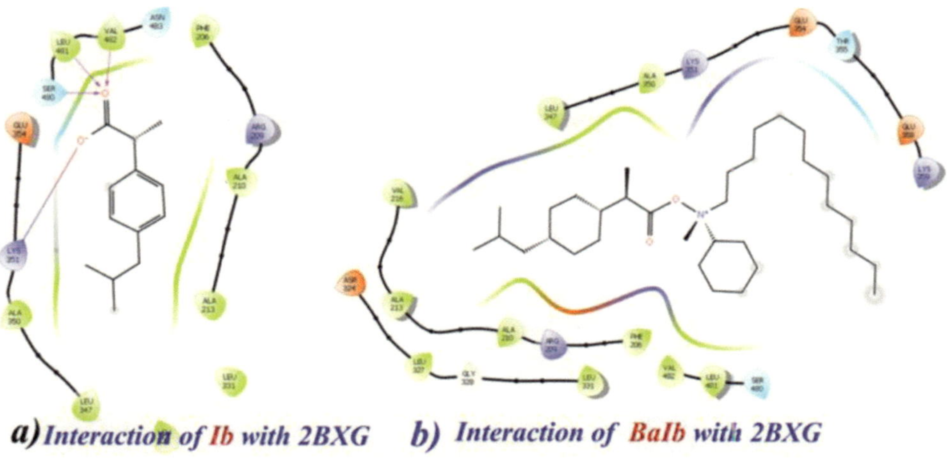

Figure 5.
Schematic representations of ligand-protein interaction and binding interaction using stick mode. (a) Ibuprofen with 2BXG and (b) BaIb with 2BXG using Schrodinger software.

HSA protein as similar to the native ligand ibuprofen by forming a hydrogen bond with active site amino acid.

4. Conclusions

In this work, the biological evaluations of a synthesized double active pharmaceutical ingredient BaIb were done to confirm the retainity of the biological activities of its parent drugs and to elucidate information regarding its potential activities against their respective ailments. Further molecular docking studies were done to get a better understanding about the mode of interaction of the parent as well as daughter drugs with the targeted proteins and trace out the binding pore and cites in the targeted proteins.

The in vitro studies revealed that the synthesized BaIb could be designed as a potential double active drug since it retained the antibacterial activity of its parent BaCl with considerable inhibitory action against *E. coli* and *P. aeruginosa* compared to the parent drug. The binding energy and docking score of BaCl and BaIb again confirm that the prepared d-API BaIb docks well into the LpxC proteins of *E. coli* and *P. aeruginosa* with high docking and Glide score compared to the parent drug BaCl.

Similarly, the results from both in vitro and in silico method emphasize that the prepared d-API retained the anti-inflammatory action of its parent NaIb and bound well to the deep pocket of the active site in the human serum albumin. Thus, in total, one can conclude that the prepared BaIb can be used as a potential double active drug with antibacterial and anti-inflammatory actions.

Acknowledgements

The authors thankfully acknowledge the fruitful discussions they had with Dr. Deshpande. The authors also acknowledge the Central Sophisticated Instrumentation Facility (CSIF), University of Calicut, for the Schrodinger software support. KPSH acknowledges UGC-MANF for fellowship with sanction number MANF2017-18-KER-78598. KPSH and MST gratefully acknowledge the collaborative research grant from UGC-DAE (No. UDCSR/MUM/AO/CRS-M-210/2015/501 dated 06/01/2015). MST further acknowledges the financial assistance from KSCSTE (SRS, SARD), UGC (MRP), and DST FIST.

Author details

Kodakkat Parambil Safna Hussan[1], Mohamed Shahin Thayyil[1], Thaikadan Shameera Ahamed[2] and Karuvanthodi Muraleedharan[2*]

1 Department of Physics, University of Calicut, Malappuram, Kerala, India

2 Department of Chemistry, University of Calicut, Malappuram, Kerala, India

*Address all correspondence to: kmuralika@gmail.com

References

[1] Sekhon BS. Ionic liquids: Pharmaceutical and biotechnological applications. Asian Journal of Pharmaceutical and Biological Research. 2011;**1**:395-411

[2] Hough WL, Smiglak M, Rodriguez H, Swatloski RP, Spear SK, Daly DT, et al. The third evolution of ionic liquids: Active pharmaceutical ingredients. New Journal of Chemistry. 2007;**31**:1429-1436

[3] Ferraz R, Branco LC, Prudêncio C, Noronha JP, Petrovski Ž. Ionic liquids as active pharmaceutical ingredients. ChemMedChem. 2011;**6**:975-985

[4] Safna Hussan KP, Mohamed Shahin T, Deshpande SK, Jinitha TV, Rajan VK, Ngai KL. Synthesis and molecular dynamics of double active pharmaceutical ingredient-benzalkonium ibuprofenate. Journal of Molecular Liquids. 2016;**223**:1333-1339

[5] Safna Hussan KP, Mohamed Shahin T, Rajan VK, Muraleedharan K. Experimental and density functional theory studies on benzalkonium ibuprofenate, a double active pharmaceutical ingredient. Computational Biology and Chemistry. 2018;**72**:113-121

[6] Hough WL, Rogers RD. Ionic liquids then and now: From solvents to materials to active pharmaceutical ingredients. Bulletin of the Chemical Society of Japan. 2007;**80**:2262-2269

[7] Oh D, Sun J, Nasrolahi Shirazi A, LaPlante KL, Rowley DC, Parang K. Antibacterial activities of amphiphilic cyclic cell-penetrating peptides against multidrug resistant pathogens. Molecular Pharmaceutics. 2014;**11**:3528-3536

[8] Gao X, Wang W, Wei S, Li W. Review of pharmacological effects of Glycyrrhiza radix and its bioactive compounds. China Journal of Chinese Materia Medica. 2009;**21**

[9] Bolton EE, Wang Y, Thiessen PA, Bryant SH. Chapter 12 - PubChem: Integrated platform of small molecules and biological activities. In: RAW, DCSBT, editors. Annual Reports in Computational Chemistry. Elsevier; 2008. pp. 217-241

[10] Lewars E. Computational Chemistry; Introduction to the Theory and Applications of Molecular and Quantum Mechanics. Netherlands: Springer Publishers; 2011

[11] Frisch MJ, Trucks GW, Schlegel HB, Scuseria GE, Robb MA, Cheeseman JR, et al. GAUSSIAN 09 (Revision A.2), Gaussian 09, B.01. Wallingford, CT: Gaussian, Inc.; 1998

[12] wwPDB consortium. 2019. Available from: www.wwpdb.org

[13] Zhou Z, Felts AK, Friesner RA, Levy RM. Comparative performance of several flexible docking programs and scoring functions: Enrichment studies for a diverse set of pharmaceutically relevant targets. Journal of Chemical Information and Modeling. 2007;**47**:1599-1608

[14] Repasky MP, Murphy RB, Banks JL, Greenwood JR, Tubert-Brohman I, Bhat S, et al. Docking performance of the glide program as evaluated on the Astex and DUD datasets: A complete set of glide SP results and selected results for a new scoring function integrating WaterMap and glide. Journal of Computer-Aided Molecular Design. 2012;**26**:787-799

[15] Warren GL, Andrews CW, Capelli A-M, Clarke B, LaLonde J, Lambert MH, et al. A critical assessment of docking programs and scoring functions.

Journal of Medicinal Chemistry. 2006;**49**:5912-5931

[16] Schulz-Gasch T, Stahl M. Binding site characteristics in structure-based virtual screening: Evaluation of current docking tools. Journal of Molecular Modeling. 2003;**9**:47-57

[17] Pagadala NS, Syed K, Tuszynski J. Software for molecular docking: A review. Biophysical Reviews. 2017;**9**:91-102

[18] Castro-Alvarez A, Costa AM, Vilarrasa J. The performance of several docking programs at reproducing protein-macrolide-like crystal structures. Molecules. 2017;**22**:136

[19] Friesner RA, Banks JL, Murphy RB, Halgren TA, Klicic JJ, Mainz DT, et al. Glide: A new approach for rapid, accurate docking and scoring. 1. Method and assessment of docking accuracy. Journal of Medicinal Chemistry. 2004;**47**:1739-1749

[20] Halgren TA, Murphy RB, Friesner RA, Beard HS, Frye LL, Pollard WT, et al. Glide: A new approach for rapid, accurate docking and scoring. 2. Enrichment factors in database screening. Journal of Medicinal Chemistry. 2014;**47**:1750-1759

[21] Friesner R, Murphy R, Repasky M, Frye L, Greenwood J, Halgren T, et al. Extra precision glide: Docking and scoring incorporating a model of hydrophobic enclosure for protein-ligand complexes. Journal of Medicinal Chemistry. 2006;**49**:6177-6196

[22] Sastry G, Adzhigirey M, Day T, Annabhimoju R, Sherman W. Protein and ligand preparation: Parameters, protocols, and influence on virtual screening enrichments. Journal of Computer-Aided Molecular Design. 2013;**27**:221-234

[23] Epik. Schrödinger Suite 2019-2 Protein Preparation Wizard; 2019

[24] Malizia WF, Gangarosa EJ, Goley AF. Benzalkonium chloride as a source of infection. New England Journal of Medicine. 2010;**263**:800-802

[25] Prasad NK. Synthesis, characterisation and biological activity of metal complexes of furoic acid. International Journal of Basic and Applied Chemical Sciences. 2015;5:52-57

Chapter 6

Hydrazone-Based Small-Molecule Chemosensors

Thiago Moreira Pereira and Arthur Eugen Kümmerle

Abstract

The hydrazone functional group is widely applied in several fields. The versatility and large use of this chemotype are attributed to its easy and straightforward synthesis and unique structural characteristics which is useful for different chemical and biological purposes. Recently hydrazone scaffold has been widely adopted in the design of small-molecule fluorescent and colorimetric chemosensors for detecting metals and anions because of its corresponding non-covalent interactions. This chapter provides an overview of hydrazone-based fluorescent and colorimetric chemosensors for anions and metals of biological interest, with their representative rational designs in the last 15 years. We hope this chapter inspires the development of novel and powerful fluorescent and colorimetric chemosensors for a broad range of applications.

Keywords: hydrazone, cyanide, acetate, fluoride, zinc, copper, aluminum, magnesium, mercury, coumarin, fluorescein, rhodamine, Schiff base

1. Introduction

Hydrazone-based molecular structures are ubiquitous in many research fields, such as medicinal chemistry [1], organic synthesis [2], supramolecular chemistry [3], metal-organic coordination [4], dyes [5], fluorescent sensors, and molecular machines [6], besides others applications [7]. Over the last decades, the popularity of hydrazone group has increased due to its easy and direct-obtaining synthesis, stability toward hydrolysis in comparison with imines, modularity, and mainly, functional diversity of C=N–N useful in several fields (**Figure 1**). In terms of structure, hydrazones are considered as azomethine compounds; however they are distinguished from imines and oximes, for example, by the presence of additional linked nitrogen atom [8]. Hydrazone backbone has an imine carbon that has an

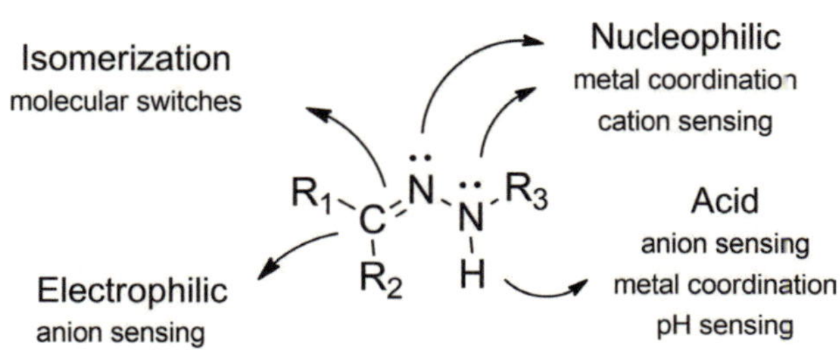

Figure 1.
General molecular structure of hydrazones.

electrophile character, two nucleophilic nitrogen in both imine and amine groups and a possible isomerization of C=N double bond typically from the conjugation of imine and the acid N–H. These structural properties play a crucial role to determine the specificity of applications which hydrazone group can be involved [6, 9].

The main synthesis of hydrazones is carried out from acid-catalyzed condensation between hydrazines (R_1NHNH_2) and activated carbonyl aldehydes or ketones, generally in alcoholic media. Other forms to obtain hydrazones are from Japp-Klingemann reaction (i.e., aryl diazonium salts coupling with β-keto esters or acids) and coupling between aryl halides and non-substituted hydrazones [9].

1.1 Hydrazone-based compounds as fluorescent chemosensors

Most hydrazone derivative fluorescent chemosensors were designed combining fluorophores or aromatic structures with this functional group. The wide range of chemical reactivity of hydrazones allows their application in the detection of anions, cations, and other species [10].

Some hydrazone-based chemosensors have weak fluorescence because of quenching effects such as E/Z double bond isomerization in the excited state; photoinduced electron transfer (PET) process (excited electron is transferred from donor to acceptor; generating a charge separation, i.e., redox reaction takes place in excited state); [11] excited state intramolecular proton transfer (ESIPT) process (photoexcited molecule relax their energy through tautomerization by transfer a proton); and others [12, 13]. The main objective for this class of chemosensors is inhibiting the quenching effects after interaction with some analytes promoting a fluorescence state. Other possible mechanisms are based on nucleophilic addition or induced N–H and O–H deprotonation. These mechanisms will be detailed after.

A quick literature survey using Scopus database has shown few reviews on the chemistry of hydrazones, most of them focusing on medicinal chemistry [14] or organic synthesis [15]. Only one review on hydrazone compounds describes some examples of hydrazone-based fluorescent chemosensor, which covered some results reported before 2014 [6].

This chapter book aims to present the progress of fluorescent and colorimetric chemosensor based on hydrazone scaffold, as reported in the literature in the period of 2006 until 2019. We hope that this chapter book helps in the design and development of new and selective fluorescent and colorimetric chemosensors for a broad range of applications.

2. Fluorescent chemosensors for anions

Anions, such as cyanide (CN^-), fluoride (F^-), chlorine (Cl^-), and acetate (AcO^-), play an important role in many environmental, clinical, chemical, and biological processes. Due to these important roles, anion recognition is an area with growing interest in supramolecular chemistry, and considerable efforts has been focused on the design of receptors (compounds) that are able to recognize anions. The detection and quantification of anions is a challenge, especially in biological systems. Some aspects as microenvironmental sensitivity, specificity, basicity, and nucleophilicity are among the main complicating factors in the detection of anions. One way to solve these problems is to develop chemosensors with high specificity for individual anions [16, 17]. Among different types of anions, fluoride and cyanide aroused great interest. The optimum concentration of F^- anions in the human body is a positive aspect to our health and can prevent dental caries and osteoporosis; however the excess of F^- may cause dental or osseous fluorosis, thyroid and liver damage, and bone diseases [18–20]. Additionally, F^- is known as a test index for residues of some nerve agents, being also associated with certain drugs, Alzheimer disease, and drinking water. The level of F^- recommended in potable water by the US Environmental Protection Agency (EPA) is about 2 ppm [18–21].

Commonly involved in chemicals and industrial processes, cyanide is highly toxic and its exposure to live organisms and environment is extremely detrimental. Moreover, cyanide anion has a strong affinity with cytochrome a_3 which can lead to cell death because of respiratory arrest [22]. According to the World Health Organization (WHO), the permissible level of cyanide in drinking water is 1.9 × 10^{-6} M [18]. Therefore, considering the notorious toxicity of CN^- and F^-, the development of sensitive sensors for the accurate detection and quantification of anions is of great importance.

2.1 Hydrazone derivatives as chemosensors for CN^-

The main mechanisms of fluorescent sensing CN^- in hydrazone derivatives are based on nucleophilic addition to polarized C=N [23, 24] and C=O [25] bonds, which leads to disruption of C=N and C=O double bond to C–NH or C–OH forms, deprotonation of NH or OH by means of acid-base reactions [26–28], and the displacement of fluorescent hydrazones from hydrazone-copper complexes.

2.1.1 Nucleophilic addition of CN^-

Two highly selective CN^- chemosensors **1–2** based on hydrazones functionalized with salicylaldehyde were described as capable of detecting this anion in aqueous solution at very low concentrations (**Figure 2A**). The ability of **1–2** complexing with several anions was tested by means of UV-vis absorption and fluorescence spectrometry. Among these anions, only CN^- caused spectral changes due to its nucleophilic attack to the imine group (**Figure 2B**), and spectroscopy analysis (H^1 NMR and MS studies) confirmed a 1:1 binding stoichiometry. In the presence of CN^- (0–120 equivalents), a new intramolecular hydrogen bond network is formed, resulting in a turn-on fluorescence response for probe **1** and colorimetric naked-eye

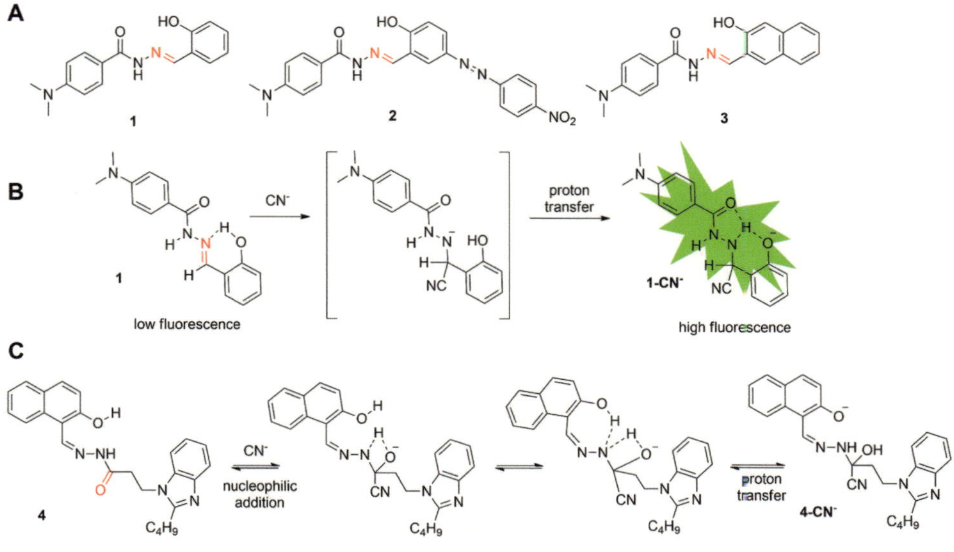

Figure 2.
*(A) Molecular structures of CN^- chemosensors **1–3** with polarized C=N bond as sensing sites. (B) Proposed cyanide sensing mechanism of **1**. (C) Proposed CN^- sensing mechanism for **4** based on polarized C=O bond.*

changes for probe **2** [23]. With the same sensing mechanism based on nucleophilic attack of CN^- on the imine group, a remarkably and selective fluorescent and colorimetric chemosensor **3** was reported (**Figure 2A**). Among various anions, sensor **3** responded to only CN^-, resulting in a color change from colorless to yellow. Moreover, a fluorescence analysis showed **3** has a weak fluorescence, and after addition of CN^- (0–120 equivalents), the fluorescence emission has increased to a bright green fluorescence [24].

Exploring the same principle of nucleophilic attack, a highly selective and sensitive naphthalene-acylhydrazone chemosensor for CN^- in aqueous media was designed. By this time, the mechanism proposed was the nucleophilic attack on the carbonyl group instead of the imine one, according to ^{1}H NMR, ^{13}C NMR, ESI-MS, and DFT calculations data (**Figure 2C**). Among several anions tested, only CN^- could induce a remarkable color change from colorless to yellow and increase of fluorescence emission in DMSO/H_2O solution. Moreover, the detection limits were 5.0×10^{-7} M and 2.0×10^{-9} M of CN^- for color and fluorescence changes respectively, far lower than the WHO guideline of 1.9×10^{-6} M [25].

2.1.2 Deprotonation mechanism

Cyanide anion is a Lewis base and can form hydrogen bonds with hydrogen bond donors as hydroxyl and amines usually followed by deprotonation.

The highly selective and sensitive chemosensor (**5**) based on acyl hydrazone could detect CN^- in aqueous solution with colorimetric and fluorimetric turn-on response (**Figure 3A**). The detection limit of CN^- was 1.2×10^{-9} M, which is lower than the maximum level of 1.9×10^{-6} M for cyanide in drinking water according to WHO guidelines. Additionally, test strips based on **5** were fabricated and demonstrated that it could be used as an efficient CN^- sensing in aqueous solution [26].

Another deprotonation mechanism was reported in the design of a two-dimensional carbazole-based chromophore **6** as chemosensor for the measurement of CN^-. In terms of structure, this compound possesses two types of donor-π-acceptor (D-π-A) chromophores, where carbazole moiety is a donor for the two branches, each one with an accepting group (**Figure 3B**). Addition of CN^- to the solution of **6** in DMSO/H_2O (95/5, v/v) redshifted from 440 nm to 500 nm its absorption, changing its naked-eye observed color from yellow to violet. The sensing mechanism proposed was based on proton abstraction of **–NH–** when this group reacts with CN^- resulting in the obvious color change [27].

Like compound **5**, a rhodamine B hydrazone derivative (**7**) was reported as highly selective chemosensor for CN^- by means of a phenol deprotonation.

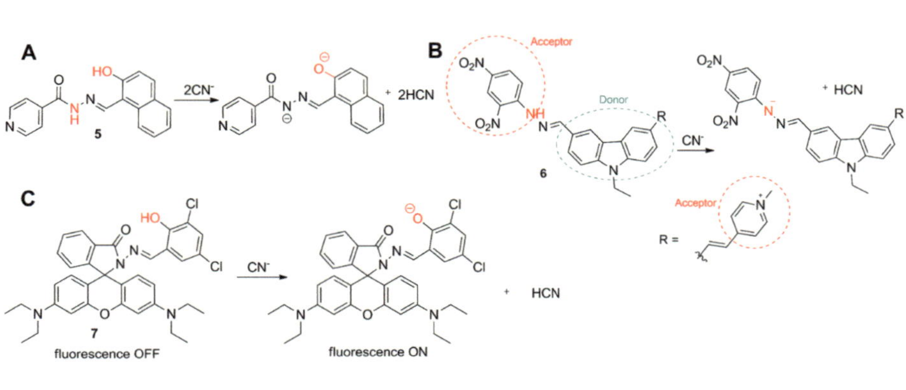

Figure 3.
Molecular structures of CN^- chemosensors 5 (A), 6 (B), and 7 (C) and their proposed sensing mechanism.

The complete sensing mechanism proposed was deprotonation followed by an intramolecular charge transfer (ICT), supported by ^{1}H NMR studies and DFT calculations (**Figure 3C**). This compound is a fluorescence and colorimetric sensor in DMSO/H_2O (1:9) medium with a significant color change after addition of CN^- to a chemosensor solution from colorless to pale yellow visible to the naked eye, and the fluorescence increased to strong green fluorescence. The good detection limit observed of 5.81×10^{-8} is lower than the maximum level of 1.9×10^{-6} M for cyanide according to WHO guidelines, leading this compound to be applied in the detection of CN^- in germinated potatoes and also in tests strips for CN^- detection [28].

2.2 Hydrazone derivatives as chemosensors for F^-

2.2.1 Hydrogen bond interactions and deprotonation

Fluoride is a weak Lewis base and can form hydrogen bonds with hydrogen bond donors. This coordination usually promotes deprotonation.

A Ru-bpy-based quinone hydrazone was designed as chromo-fluorogenic hybrid chemosensor (**8**) for F^- (**Figure 4A**). This compound contains a quinone-hydrazone group that can be converted to azophenol tautomer in **8-F**$^-$ induced by the proton transfer from **8** to F^- causing a color change from orange to blue-violet. Only F^- was capable of inducing color change to **8** in MeCN, suggesting the high selectivity could be attributed to the strong intramolecular N–H–O hydrogen bond interaction in **8**, which means only the most electronegative anion could form an additional hydrogen bond. Generally, anion sensors based on hydrogen bond interactions cannot serve as good sensors in aqueous media due to hydrogen bond competition with water. To avoid this problem, a filter paper impregnated with acetonitrile solution of **8** and dried in air has been prepared. Immersing this paper into aqueous fluoride solution was successful, and fluoride solution exhibited color changes [29].

A thiocarbonohydrazone anion chemosensor **9** was described and rationally designed based on previously reported anion chemosensors, where the presence of strong electron withdrawing $-NO_2$ group enhanced the acidity of the thioamide protons and stabilized the negative deprotonated species (**Figure 4B**). After successive addition of F^-, the UV-vis absorption band with maximum at 360 nm has decreased, whereas new peaks at 407 and 495 nm appeared. The absorption at 360 nm is attributed to the Ar–CH=N–NH conjugation moiety, and its bathochromic shift clearly indicated an interaction/reaction of fluoride with this portion. After addition of more than four equivalents of F^-, a new absorption band at 600 nm appeared with a new isosbestic point at 535 nm. The significant changes in UV-vis spectra of chemosensor **9** was attributed to the deprotonation of thioamide protons in a three-step process (**Figure 4B**) confirmed by ^{1}H NMR titration analysis from the observed peak of the FHF– at δ = 16.13 ppm [30].

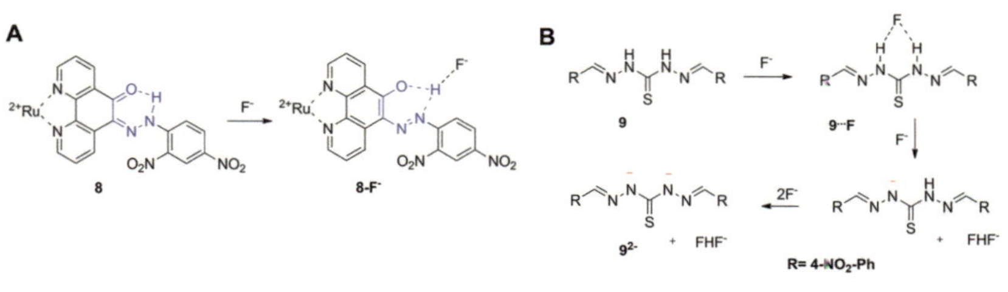

Figure 4.
Molecular structure of F⁻ chemosensor 8 (A) and 9 (B) and their proposed sensing mechanism.

Still exploring the N–H acidity of hydrazones, a new series of diketopyrrolopyrrole (DPP) derivatives **10–14** bearing phenylhydrazone group was described presenting a ESIPT-PET fluoride sensing mechanism. The authors aimed to study the effect of nitro substituent of phenylhydrazone on their photophysical property and optical fluoride sensing. The anion sensing capabilities of **10–14** were evaluated in DMSO by the addition of several anions. The presence of nitro substituent at *ortho*-position of phenylhydrazone in **11** and **14** significantly altered the electronic properties through intramolecular hydrogen bonding and furnished an excited state intramolecular proton transfer (**Figure 5**). So, *o*-NO_2-DPPH (**11**) has a weak fluorescence, partly attributed to the photoinduced electron transfer from the imine and similar situation was found in *o, p*-NO_2-DPPH (**14**) (**Figure 5**).

The visual naked-eye color change was observed under natural light for DPPPH (**10**), *m*-NO_2-DPPPH (**12**), *p*-NO_2-DPPPH (**13**), and *o, p*-NO_2-DPPPH (**14**) in the presence of F^-. Among these designed chemoreceptors, *p*-NO_2-DPPPH demonstrated selectivity for F^- in UV-vis and fluorescence evaluations in DMSO solution. Additionally, experimental results of ^1H NMR and ^{19}F NMR revealed that the spectral changes occur due to deprotonation of the hydrazone N–H moiety by fluoride ion [31].

2.3 Hydrazone derivatives as chemosensor for AcO^-

Acetate (AcO^-) and dicarboxylate are essential components in several metabolic processes in living organisms. Without them, many enzymes and antibodies are unable to function properly. In this sense, the synthesis of chemosensors that can recognize AcO^-, mainly via hydrogen bond interaction, is of great importance for biological systems [32].

An interesting naked-eye selective colorimetric sensor for AcO^- based on 1,10-phenanthroline-2,9-dicarboxyaldehyde-di-(p-nitrophenylhydrazone) (**15**) was described by Lin's group. The UV-vis absorption in DMSO showed a dramatic color change from yellow to green in the presence of AcO^- with no changes for other anions. The presence of electron withdrawing groups increasing the hydrogen bond donor ability of N–H framework was favorable for AcO^- sensing. In addition, ^1H NMR studies showed that after interaction, AcO^- lead to deprotonation of **15** [33].

Similarly a fluorescent and naked-eye colorimetric chemosensor (**16**) for AcO^- based on thiosemicarbazone was evaluated by UV–vis spectroscopic titrations in dry DMSO solution, presenting high selectivity and affinity for AcO^-. After addition of this anion, the absorption band at 373 nm decreased gradually, whereas a new band appeared at 457 nm, followed by color change in the solution from light yellow to orange. The complex formed between AcO^- and **16**, through hydrogen bond interactions, caused intramolecular charge transfer between the electron-rich urea unit and the electron-deficient benzene moiety (**Figure 6**). In addition, the fluorescence spectroscopic titrations were carried out, and the **16** also displayed a switch-on reaction toward AcO^-, which was attributed to the binding-induced

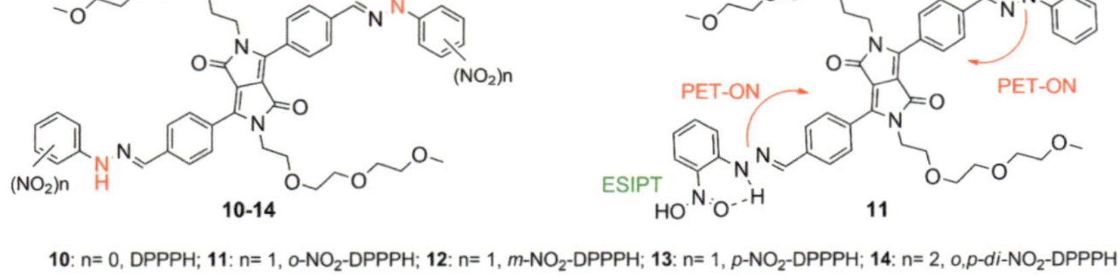

Figure 5.
*Molecular structures of F^- chemosensors **10–14**.*

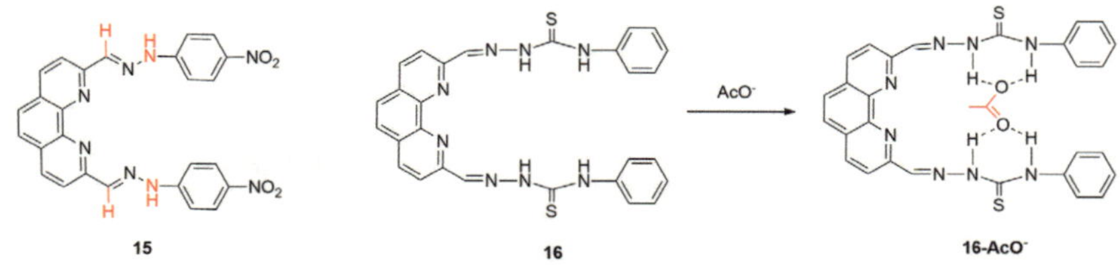

Figure 6.
*Molecular structures of acetate chemosensors **15–16** and the proposed sensing mechanism.*

rigidity of the host molecule. The free receptor (**16**) had a flexible configuration and could rotate freely; however, after complexation with AcO^-, the host molecule was rigidified, and the fluorescence emission has increased [32].

2.4 Hydrazone derivatives as chemosensors for multiple anions

In contrast to previously described, where receptors had a certain degree of selectivity toward single anion, several hydrazone-based chemosensors have been published exhibiting a response to more than one anion species, and some interesting examples will be described below [34–36].

A tripodal benzaldehyde-phenylhydrazone (**17**) was developed as a colorimetric naked-eye chemosensor for AcO^-, F^-, and $H_2PO_4^-$ by Lin and colleagues, and UV-vis spectroscopic studies in DMSO were used to determine the binding mode of **17**. After addition of these anions, a naked-eye color change from yellow to purple and a significant bathochromic shift of 124 nm were observed. However, both the color and spectral changes were reverted by the addition of protic solvent such as H_2O into the mentioned host-guest system, indicating that protic solvents could compete with anion binding sites. Considering these results, the authors proposed that a strong hydrogen bonding interaction was taking place between receptor **17** and AcO^- as well as F^-, OH^-, and $H_2PO_4^-$. The complete interaction mode described in **Figure 7** was only attributed after assays indicating that stoichiometry of the host and specific guests was different depending on the anion. The receptor formed 1:1 complex with F^- ion and 1:3 complexes with AcO^-, OH^-, and $H_2PO_4^-$. This was attributed to the smaller ionic radius of F^- than that of other larger ones. A further ^1H NMR investigation showed that resonance peak at 11.77 ppm, attributed to NH protons, exhibited a downfield to 12.27 ppm upon addition of 2 equivalents of F^- ion confirming the formation of $NH^{\cdots}F^-$ hydrogen bonding [34].

A Schiff-base thiophene-based hydrazone (**18**) was described as visual anion chemosensor in aqueous media exhibiting sensing properties for F^-, AcO^-, and $H_2PO_4^-$, among several anions tested with colorimetric response changes from

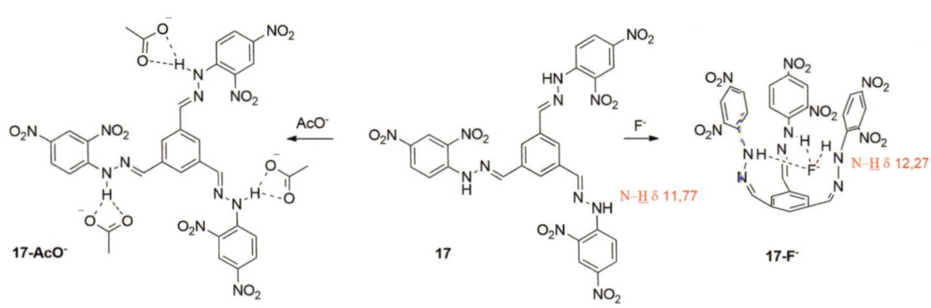

Figure 7.
*Molecular structures of multianalyte chemosensor **17** and their proposed sensing mechanism.*

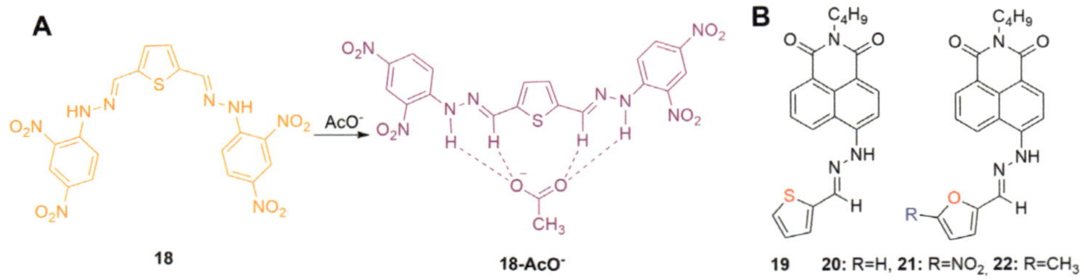

Figure 8.
*Molecular structure of multianalyte chemosensor **18** (A) and **19–22** (B) and mechanism of acetate detection for **18** (A).*

orange to violet at a micromolar level. The UV-vis analysis confirmed the naked-eye colorimetric changes and showed a decrease in band centered at 420 nm and increase in intensity of the band at 590 nm with the clear isosbestic point at 520 nm after addition of F^-, AcO^-, and $H_2PO_4^-$ to **18** solution. The chemosensor **18** works by means of a N–H deprotonation mechanism. ^1H NMR analysis confirmed this mechanism, showing that **18** presented signals at δ 11.79 and 8.91 corresponding to N–H and imine protons, respectively. After addition of AcO^- and F^- ion, N–H signal disappeared (deprotonation), while the signal corresponding to the imine and phenyl rings shifted to the upfield at the region of δ 8.59 and 8.54 (**Figure 8**). Additionally, a real sample qualitative estimation analysis of F^- and AcO^- in commercially available toothpaste and vinegar was successfully achieved by this simple and easy colorimetric method [35].

Four furan/thiophene-based fluorescent hydrazones **19–22** were described as CN^- and F^- sensors (**Figure 8**) and could detect these ions with naked-eye color changes from yellow to blue, while their fluorescence emission intensities were completely quenched. The presence of electron donating/withdrawing groups attached to furan ring as in **21** ($-NO_2$) and **22** ($-CH_3$) curiously resulted in increased selectivity for CN^- ions compared to F^- ones. ^1H NMR confirmed that the sensing mechanism goes through hydrogen bonding interaction between sensors and F^-/CN^-, followed by deprotonation, leading to elicited ICT. Job's plot afforded a stoichiometry of 2:1 binding ratio between **19** and **20** and F^- ions. However, curiously **21** and **22** exhibited a 1:1 ratio with F^- and CN^- due to steric constraint. The limit of detection (LOD) analysis revealed that the four sensors displayed the LOD below 0.3 ppm for CN^- and F^- and a good selectivity. Competitive experiments revealed a negligible perturbation in the optical response which confirms a higher selectivity for F^- and CN^- than other competitor anions [36].

3. Fluorescent chemosensors for metal ions

Metal ions such as Cu^{2+}, Zn^{2+}, Fe^{3+}, Al^{3+}, Hg^{2+}, Mg^{2+}, etc. play an important role in many biological and environmental processes, and excessive or insufficient amounts may lead to diseases [37]. As an example, copper (Cu^{2+}) is the third most abundant transition metal in the human body and plays essential roles in several environmental, chemical, and physiological systems. In living organisms, Cu^{2+} acts as a key catalytic center in many enzymes and as cofactor in a variety of metalloproteins [38]. Its insufficient concentration may affect the development of bone and brain tissues as well as the nervous and immune system, whereas excessive intake may lead to serious problems including cirrhosis and neurological diseases such as Alzheimer's and Wilson's diseases and prion disorders [39]. The extreme toxicity of heavy metal ions such as Pb^{2+} and Hg^{2+}, even in small amounts, remains a danger to human health and the environment, but they have been widely used in industrial processes [40]. Therefore, the development of sensitive sensors for the accurate detection and quantification of these ions is of great importance.

3.1 Hydrazone derivatives as chemosensors for Cu^{2+}

Fluorescent and colorimetric hydrazone-based chemosensors for Cu^{2+} attract interest and are mainly based on coordination mechanism, often quenching the fluorescence emission due to PET mechanism [41].

Coumarins are widely associated with hydrazones for sensing Cu^{2+} and the on–off fluorescent chemosensor (**23**) was described for Cu^{2+} detection in aqueous media. This chemosensor showed very strong luminescence in H_2O/DMSO (9:1, v/v) with quantum yield of 0.289, which was almost completely quenched after addition of copper (1 equivalent), decreasing the quantum yield to 0.024. This process was associated with the complexation of Cu^{2+} to the tautomeric enol-like form of **23** leading to **23-Cu^{2+}** and the PET mechanism (**Figure 9**). Compound **20** showed detection limit of 0.1 μM for Cu^{2+}, which is useful to sense Cu^{2+} in blood system, a 1:1 binding mode supported by a Job's plot, an association constant estimated to be $6.4 \times 10^5\ M^{-1}$, and a response time of 2 min upon addition of 1 equivalent of this cation. Only Cu^{2+} causes a significant fluorescence decrease, while other metals such as K^+, Ag^+, Ca^{2+}, Cd^{2+}, Co^{2+}, Cr^{3+}, Fe^{2+}, Fe^{3+}, Hg^{2+}, Mg^{2+}, Mn^{2+}, Ni^{2+} and Zn^{2+} did not cause any significant change. In addition, living cell experiments were successfully applied, and confocal image changes were observed, demonstrating its value in practical applications and biological systems [42].

Differently from turn-off PET mechanism, off-on (turn-on) chemosensors for detecting Cu^{2+} often present a FRET mechanism (described by energy transfer between two light-sensitive molecules or chromophores, where a donor chromophore in its electronic excited state may transfer energy to an acceptor chromophore through nonradiative dipole-dipole coupling). An example is the hybrid coumarin-rhodamine hydrazone chemosensor **24** based on metal ion-induced FRET. In this mechanism, coumarin nucleus acts as donor and rhodamine acts as acceptor of energy. Free ligand coumarin-rhodamine hydrazone (**24**) absorbs around 460 nm, which is attributed to coumarin chromophore. The absorbance remained unchanged upon addition of various metal ions except Cu^{2+}. Upon addition of Cu^{2+}, the solution color changes from yellow to bright red indicating metal complexation. It was confirmed analyzing the decrease in absorption band centered at 460 nm with a slowly redshifts to 475 nm, while a new peak at 556 nm arise from the rhodamine chromophore in the visible region.

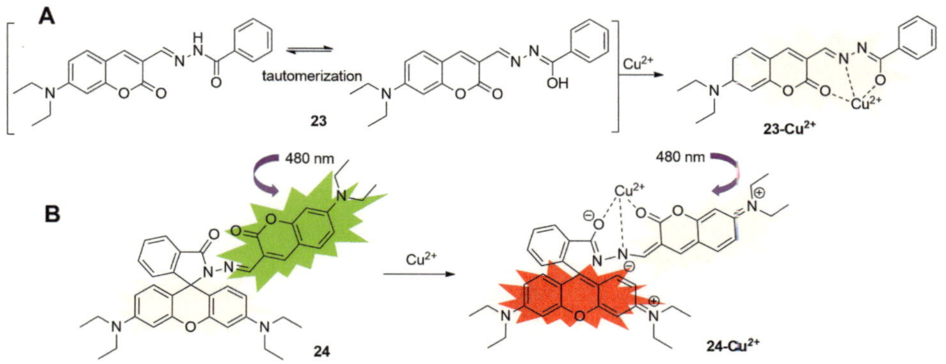

Figure 9.
*Molecular structures of Cu^{2+} chemosensors **23** (A) and **24** (B) and their proposed sensing mechanism.*

Free coumarin-rhodamine hydrazone (**24**) emits in the green fluorescence region with a fluorescence band centered at 524 nm which is attributed to *N,N*-diethyl-coumarin moiety upon excitation at 480 nm. Upon sequential addition of Cu^{2+}, coumarin emission band was observed at 524 nm, while an emission signal corresponding to ring-opened rhodamine appeared at 582 nm (**Figure 9**). This chromo-fluorogenic probe can detect concentrations below 20 μM of Cu^{2+} in aqueous buffer medium. The FRET mechanism is possible due to integral overlap between emission band of *N,N*-diethyl coumarin moiety and absorption band of ring-opened zwitterionic rhodamine unit in buffer medium. Additionally, probe **24** undergoes a 1:1 stoichiometric complexation with Cu^{2+} with the calculated association constants of 8.81 M^{-1} and was successfully employed as ratiometric biosensor for living cell imaging of Cu^{2+} [43].

A highly selective and sensitive naked-eye colorimetric chemosensor for Cu^{2+} in aqueous solution was designed and developed based on hydrazone framework. In the UV-vis spectroscopic studies, compound **25** exhibited a broad band at 336 nm, and after addition of copper, a new absorption band at 502 nm appeared, whereas the absorption band at 336 nm was gradually reduced. This gradually increasing absorption peak at 502 nm (after sequential addition of Cu^{2+}) was attributed to the coordination of **25** with Cu^{2+}.

Trying to understand its sensing mechanism, the stoichiometry of the **25-Cu2+** complexation was determined by the Job's plot analysis which indicates a 2:1 stoichiometric between **25** and Cu^{2+} and after confirmed by ESI/MS analysis. With stoichiometry on hands, complementary infrared (IR) and ^{1}H NMR spectroscopy were employed, and even Cu^{2+} which is a paramagnetic ion helped to describe the coordination mode. The ^{1}H NMR revealed that N–H (**a**) proton almost completely disappeared upon addition of Cu^{2+} and proton **b** became broader after binding, indicating that Cu^{2+} binds with nitrogen of pyridine after an initial tautomerization with a deprotonation of O–H.

The selectivity of **25** over other metals was investigated by adding several metal cations to the solution of **25** in THF/H_2O (9:1, v/v), and there was no obvious change with any other metal. The colorless solution of the compound **25** became pink after addition of Cu^{2+} with the detection limit by the naked eye around 2 μM which is lower than the limit of copper in drinking water (~20 μM). Additionally, to evaluate the practical application of chemosensor **25**, competition experiments of Cu^{2+} mixed with other metal ions were carried out from UV-vis absorption spectra. The treatment of **25** solution with Cu^{2+} in the presence of the same concentration of other metal cations did not show any significant changes [44].

Using a strategy of intraligand charge transfer transition (ILCT) turn-on mechanism, a small chromo-fluorogenic chemosensor (**26**) for Cu^{2+}, based on hydrazones, was described. This chemosensor exhibited a weak fluorescence in DMSO with a 44-fold increase in fluorescence emission intensity upon addition of Cu^{2+}, being attributed to ILCT. This receptor showed a prominent colorimetric change from yellow to brown in the presence of Cu^{2+} with a detection limit in the order of 10^{-8} M. In the presence of other environmentally significant metal cations such as Hg^{2+}, Pb^{2+}, Cd^{2+}, Ni^{2+}, Co^{2+}, Fe^{2+}, Fe^{3+}, Mn^{2+}, Zn^{2+}, Al^{3+}, and Cr^{3+}, no significant spectral changes were observed. Interesting, this chemosensor was capable of extracting Cu^{2+} selectively from aqueous mixture of metal ions using dichloromethane as solvent with an efficiency of 94% at 6.5–11 pH range.

The binding sense mechanism was explained by ^{1}H NMR titration in DMSO, in which the peak attributed to acid O–H proton of **26** gradually decreased upon addition of Cu^{2+}, also indicating a 2:1 (**26-Cu^{2+}**) stoichiometric binding (**Figure 10**). During the extraction process, the stoichiometry of ligand and Cu^{2+} was confirmed to be 2:1, and, in addition, the color changes and concentration of Cu^{2+} were

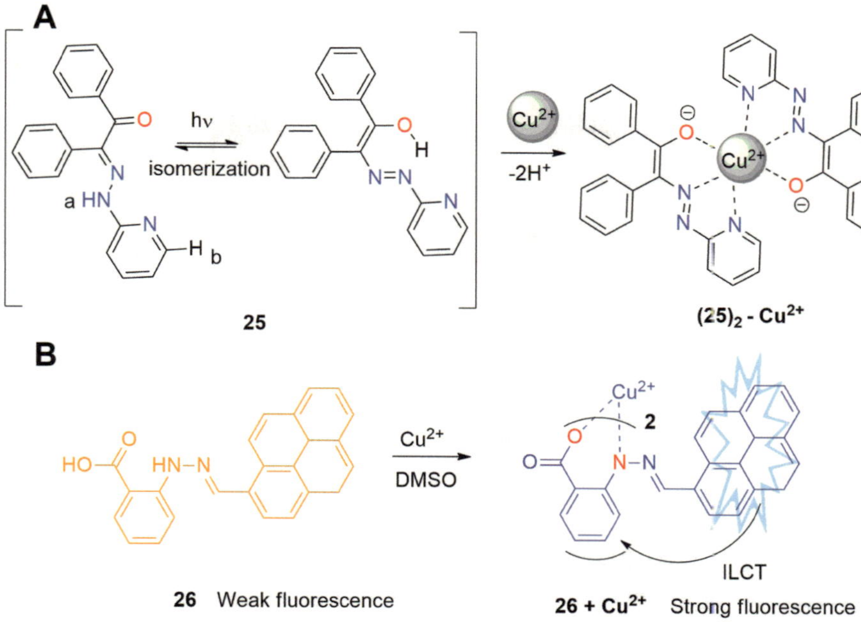

Figure 10.
Molecular structures of Cu^{2+} chemosensor **25** *and its coordination mode (A) and structure of chemosensor* **26** *and its sensing mechanism (B).*

monitored by a readily usable smartphone as an analysis tool. Interesting, this receptor showed a good recyclability and reusability in Cu^{2+} extraction, being very useful for the detection and selective extraction of Cu^{2+} from aqueous media in chemical and biological systems [45].

3.2 Hydrazone derivatives as chemosensors for Zn^{2+}

Zinc (Zn^{2+}) is the second most abundant transition metal (after Fe^{3+}) in the human body and is considered essential for living organisms. Zn^{2+} exerts influence on many cellular processes, including proliferation, differentiation, apoptosis, transcription, neural signal transmission, and microtubule polymerization. Therefore, significant changes in Zn^{2+} concentration may be related to many diseases, including Alzheimer's and Parkinson's diseases, diabetes, and prostate cancer [46, 47]. Chemosensors for Zn^{2+} are mainly based on the coordination mechanism; however, these probes often still lack specificity.

Aroylhydrazone derivatives (**27–28**) were described as fluorescent chemosensors for Zn^{2+} recognition. These ligands and their metal complexes (**27** (Zn^{2+}, Cu^{2+}), **28** (Zn^{2+}, Co^{2+}, Fe^{3+})) have been synthesized and characterized in terms of their crystal structures, elemental analysis, and spectroscopic properties. First, chemosensor **27** displayed high selectivity for Zn^{2+} over other transition metals compared to **28** in aqueous ethanol solution, which indicate that hydroxyl group exerts effect on the selectivity of the fluorescent chemosensor. The possible reason is that the presence of hydroxyl group gives the carbonyl a better binding ability for Zn^{2+} to form a 1:1 complex. The fluorescence response of **27** in solution increased approximately 25-fold upon addition of 10 equivalents of Zn^{2+} probably due to rigidity imposed by the complex formed, inhibiting isomerization (**Figure 11**). Otherwise, **28** exhibited only small increase in fluorescence (5-fold) when Zn^{2+} was introduced to the solution, indicating also the importance of *ortho*-hydroxyl group for fluorescence [48].

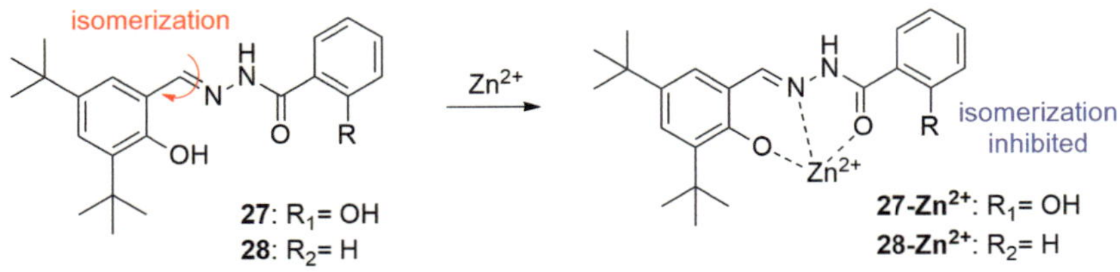

Figure 11.
*Molecular structures of Zn^{2+} chemosensors **27–28** and their sensing mechanism.*

A fluorescein-coumarin conjugate (**29**) was reported as turn-on fluorescent sensor for Zn^{2+} in aqueous medium. The mechanism involved in this chemosensor is related to spirolactam ring opening mediated selectively by Zn^{2+} over other earth and transition metal ions. The free chemosensor **29** showed almost no absorption characteristic of the fluorescein moiety which indicate the existence of the spirolactam form. However, upon Zn^{2+} addition in the sensing system, a new absorption band corresponding to fluorescein moiety increased indicating the generation of a ring-opening amide form (**29-Zn^{2+}**). The absorbance ascended linearly as a function of Zn^{2+} concentration with a saturation at the ratio of 1:1 (**Figure 12**). The fluorescence emission intensity was increased 33-fold at 501 nm upon zinc addition, which is characteristic of fluorescein, confirming the spirocycle opening of **29** after coordination. In addition, this chemosensor was highly selective toward several metals such as Na^+, K^+, Ca^{2+}, Mg^{2+}, Cu^{2+}, Fe^{3+}, Co^{2+}, Ni^{2+}, Hg^{2+}, Cd^{2+}, and Cr^{3+} that showed none or little fluorescence intensity changes. The ability of chemosensor **29** works in aqueous medium and shows potential applications in environmental, biological, and medicinal areas [49].

The hydrazone-based fluorescent chemosensor **30** was reported as an interesting and selective sensor for Zn^{2+} over other biologically important metal ions (Na^+, K^+, Mg^{2+}, Ca^{2+}, Co^{2+}, Ni^{2+}, Cu^{2+}, Cd^{2+}, Ag^+, Pb^{2+}, Hg^{2+}, Al^{3+}, Cr^{3+}, Fe^{3+}, and Zn^{2+}) under completely physiological conditions (HEPES buffer medium (1 mM, pH = 7.4) containing 0.33% of DMSO), also demonstrating detection of zinc in live Hela cells by fluorescence imaging. The recognition of Zn^{2+} was investigated by absorption and emission spectroscopy, DFT calculations, ESI-MS experiment, and ^{1}H NMR titration.

The free ligand (**30**) exhibited two absorption peaks at 305 nm and 367 nm attributed to the π-π* transitions, in which zinc addition (10 equivalents) promoted a prominent change where the absorption band at 305 nm decreased, whereas a new peak at 450 nm emerged. This absorption spectra change became minimum upon the addition of two equivalents of metal ions, which suggested a 1:2 ratio between **30** and Zn^{2+}. Free receptor (**30**) has weak fluorescence with visible light excitability; however zinc caused drastic enhancement in the fluorescence emission with a prominent peak at 555 nm. The Zn^{2+} sensing mechanism was attributed to

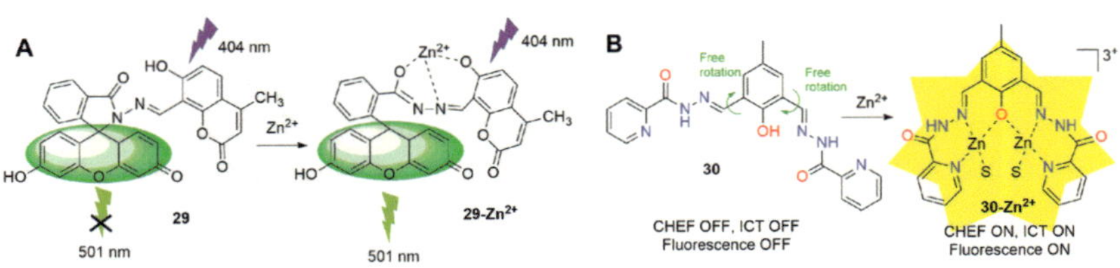

Figure 12.
*Molecular structures of Zn^{2+} chemosensors **29** (A) and **30** (B) and their sensing mechanism.*

the chelation-enhanced fluorescence (CHEF) and ICT processes. The low fluorescence state of **30** (Φ = 0.009) is may be due to free rotation of imine (**–**C=N) bond (**Figure 12**). Upon Zn^{2+} addition, the coordination of this metal with imine nitrogen and the hydroxyl group inhibited the free rotation around the imine bond, leaving the system more rigid. Thus, the chelation of Zn^{2+} leads to the formation of binuclear zinc complex with drastic increase in conjugation, resulting in a CHEF effect. Furthermore, due to the binding of Zn^{2+}, the ICT is facilitated over the π-system. These conjugated effects caused the improvement in the fluorescence emission (Φ = 0.16, 18-fold) of **30** (**Figure 12**) [50].

3.3 Hydrazone derivatives as chemosensors for Hg^{2+}

Mercuric ion (Hg^{2+}) is considered highly dangerous, because it is known as one of the most toxic metal ions and is generated by many sources such as mercury lamps, gold production, electronic equipment, paints, and batteries [51]. Mercuric ion can cause serious detrimental effects to living organisms, resulting in hepatitis, uremia, digestive diseases, and fatal damage to the central nervous system, and its accumulation can lead to various cognitive and motor disorders, such as Minamata disease [52]. Due to its high toxicity, considerable attention has been devoted to the development of new sensors for Hg^{2+} detection. Hydrazone-based fluorescence chemosensors for Hg^{2+} are mainly based on coordination mechanism.

One example is the 3,4-ethylenedioxythiophene (EDOT) rhodamine-hydrazine-based compound **31** which acts as colorimetric and turn-on fluorescent chemosensor for Hg^{2+} detection. Between several metals, only Hg^{2+} coordination promoted the turn-on effect. The sensing mechanism is quite similar to that observed to compound **29**, in which free receptor has a spirolactam moiety, which inhibits the intramolecular charge transfer between electron-acceptor moiety of xanthene and the electron donor of EDOT. Once Hg^{2+} complexation occurs, a ring opening takes place delivering the rhodamine B moiety, which is a well-known fluorophore. The spirolactam moiety of the rhodamine acts as a signal switcher, which is envisioned to turn on upon complexation with the cation. The spectroscopic parameters of Hg^{2+} in mixed ethanol and HEPES (10 mM, 1:1, v/v, pH = 7.2) solution demonstrated that only Hg^{2+} was capable to change the color of the sensing solution from colorless to red (570 nm) and to increase fluorescence emission (593 nm), also indicating a 2:1 complex formation (**Figure 13**) [53].

The previous exposed FRET mechanism has been used in the design of selective turn-on fluorescent chemosensor for Hg^{2+} based on bis-hydrazone derivative from 2,5-furancarboxaldehyde. In this case, furan ring is the donor, and rhodamine B is the acceptor chromophore. Free ligand has the spirolactam moiety which in turn inhibits the charge transfer between these chromophores. When Hg^{2+} binds to **32**, a rapid naked-eye visual color change occurs from colorless to sharp pink, as well as a bright red fluorescent emission under UV lamp irradiation, which is attributed to the spirolactam ring opening (**Figure 13**). The FRET mechanism was confirmed by overlap of the emission band of 2,5-furan-dicarboxaldehyde (donor) and absorption spectra of rhodamine B (acceptor). After the binding of **32** to Hg^{2+}, leading to **32-Hg^{2+}**, furan moiety makes an energy transfer to induce the spirolactam opening which allowed the FRET process with increasing in the conjugation of the system (**Figure 13**). Additionally, the high increase in quantum yield of **32-Hg^{2+}** (0.23) when they compared **32** (0.015) and its limit of the detection at very low ppb level concentration (3.6×10^{-9} M) allowed applications for detecting Hg^{2+} in drinking water (LOD = 10 nM) [54].

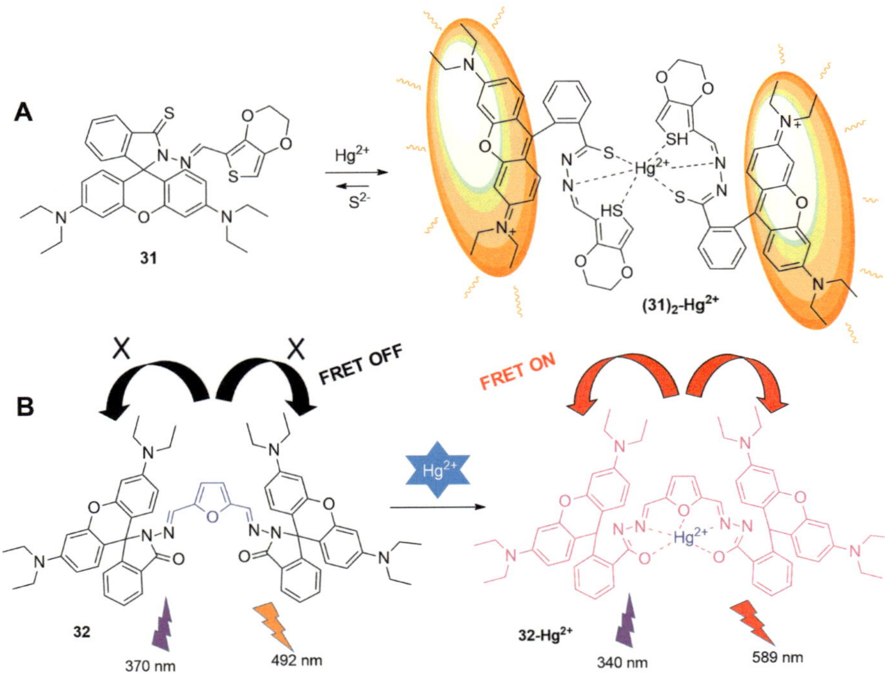

Figure 13.
Molecular structure of Zn^{2+} chemosensors 31 (A) and 32 (B) and their sensing mechanism.

3.4 Hydrazone derivatives as chemosensors for Al^{3+}

Aluminum (Al^{3+}) is the most abundant (8.3% by weight) metallic element and, after oxygen and silicon, is the third most abundant of all elements in the earth. Aluminum is widely used in the environment around us in modern society, such as in water treatment, food packing, medicines, etc. However, the excess of this metal can result in health problems such as Alzheimer's and Parkinson's diseases [55]. Moreover, it is believed that around 40% of the world's acid solid are caused by aluminum toxicity, which is harmful to plants' performance [56]. Thus, the detection of aluminum is essential in controlling its effects on environment and on human health. Hydrazone-based chemosensors for aluminum ion (Al^{3+}) are mainly based on coordination with fluorescence turn-on response as a result of restricted molecular motion through inhibiting ESIPT or PET effects.

A Schiff-base 7-methoxychromone-3-carbaldehyde-(pyridylformyl) hydrazone was reported as turn-on fluorescent and colorimetric chemosensor for Al^{3+}. This chemosensor (**33**) is colorless and nonfluorescent either in aqueous medium or organic solvents; however, in the presence of aluminum ions (Al^{3+}), the development of a yellow-green color and yellow-green fluorescence occurred. The emission intensity of **33** is very low with low fluorescence quantum yield of 0.051 in ethanol, being attributed to a PET mechanism from the Schiff-base nitrogen free pair electrons. As exposed, PET process involves the deactivation of excited state of a fluorophore by the addition of an electron to one of its excited state frontier orbitals which leaves the fluorophore in a non-emissive state.

Metal complexation to Schiff base produces a less efficient electron donor character, interrupting the PET process and, in some cases, improving the fluorescence emission, which is known as CHEF effect [57]. The selective coordination between **33** and Al^{3+} on the carbonyl of chromone, nitrogen of imine (–C=N), and carbonyl of pyridylformyl hydrazone moiety suppressed the PET effect, restoring the fluorescence of the system with an increase of more than 800-fold (**Figure 14**) [58].

The simple and selective fluorescent naphthalene-hydrazone chemosensor (**34**) for Al^{3+} is a good example of chemosensor based on the excited state intramolecular proton transfer mechanism, that is a process in which photoexcited molecules relax their energy through tautomerization by transfer of protons. Compound **34** has a characteristic UV-vis bands 325 nm and 366 nm which should be assigned to π-π^* transitions of the naphthalene. Only in the presence of Al^{3+} the spectra of **34** exhibited a peak at 432 nm, which remained constant even after more than 1 equivalent of aluminum addition, which indicates 1:1 binding stoichiometry between **34** and Al^{3+}. The free receptor (**34**) exhibited no fluorescence emission with low fluorescence quantum yield (0.5%), justified by the electron transfer from nitrogen atom of imine to the naphthalene ring (PET) and also transfer of the hydroxyl proton to a neighboring imine nitrogen along with the formation of intramolecular hydrogen bond (OH¨N) (ESIPT). Upon addition of several metal ions, only Al^{3+} could cause a significant enhancement in the fluorescence emission at 475 nm with a high fluorescence quantum yield (0.26%), due to PET and ESIPT process inhibition (**Figure 14**). Chemosensor **34** showed an interesting fluorogenic response to Al^{3+} in fully aqueous medium which allows its application in biological assays and environmental systems [55].

3.5 Hydrazone derivatives as chemosensors for multiple metals

Although several chemosensors have relatively high degree of selectivity as previously exposed, some chemosensors have been reported to recognize more than one metal ion.

Following a previous described strategy, probes based on the opening of spirolactam ring upon metal coordination were designed as single molecule multianalyte (Cu^{2+} and Hg^{2+}) chemosensors. Compound **35** was reported as colorless and nonfluorescent in aqueous or organic medium (**Figure 15**). The UV-vis spectroscopy indicated that this chemosensor is a good chromogenic probe for Cu^{2+} in ethanol-water (1:99, v/v), whereas other competitive cations failed. Upon the addition of Cu^{2+} to the solution of **35**, a strong absorption band centered at 530 nm appeared, with changes from colorless to pink, because of spirolactam opening (**35-Cu**$^{2+}$). A significant increase in the fluorescence emission in ethanol was also observed in the presence of Hg^{2+} (ϕ = 0.335) and Cr^{3+} (ϕ = 0.445). However, small addition of water to the ethanol quenched the fluorescence produced by Cr^{3+}, whereas the fluorescence intensity of **35-Hg**$^{2+}$ declined just a little. The rapid quenching of the **35-Cr**$^{3+}$ is justified due to strong coordination between Cr^{3+} and water which may lead to

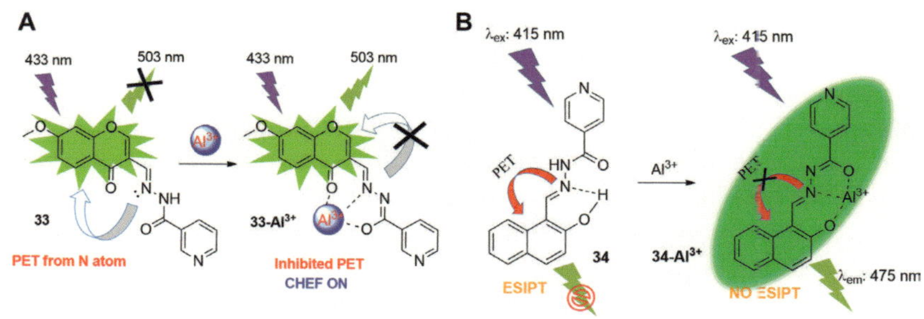

Figure 14.
Molecular structures of Al^{3+} chemosensors 33 (A) and 34 (B) and their sensing mechanism.

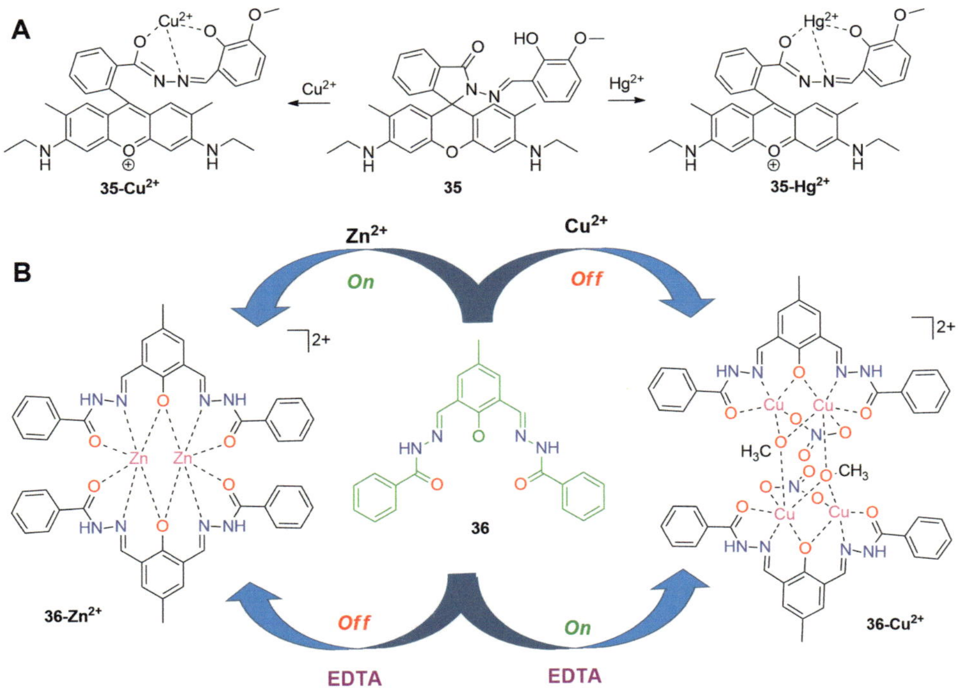

Figure 15.
Molecular structures of Cu^{2+} and Hg^{2+} multianalyte chemosensor 35 (A) and of Cu^{2+} and Zn^{2+} multianalyte chemosensor 36 (B).

hydrolysis of **35-Cr^{3+}**, resulting in $Cr(OH)_3$. In addition, the open ring form of **35** after binding with Cu^{2+} has no fluorescence, which is attributed to the quenching of the fluorescence by Cu^{2+}, due to paramagnetic properties of the d^9 Cu^{2+} system [59].

With a similar structure to compound **30**, a hydrazone-based chemosensor (**36**) with *off-on* fluorescence response to Cu^{2+} and Zn^{2+} ion in aqueous media was reported. The reaction of **36** with Cu^{2+} and Zn^{2+} formed their corresponding dimeric complexes, which were characterized by single X-ray analysis (**Figure 15**).

The UV-vis absorption and fluorescence spectroscopy (CH_3CN/0.02 M HEPES buffer at pH 7.3) indicated the binding behavior of chemosensor **36** toward Zn^{2+} and Cu^{2+}. The electronic spectra of **36** exhibit two sharp bands at 300 and 363 nm, and upon gradual addition of Cu^{2+} and Zn^{2+}, new absorption bands appeared at 423 and 415 nm, attributed to charge transfer in complexes **36-Cu^{2+}** and **36-Zn^{2+}**. The occurrence of three well-defined isosbestic points demonstrated an equilibrium between **36** and **36-M^{2+}**. The little fluorescence presented by **36** (at 493 nm) was almost quenched upon sequential addition of Cu^{2+}, being ascribed to the reverse PET from the 4-methylphenyl moiety to the phenolic hydroxyl, and carbohydrazide nitrogen and oxygen atoms, arising from the decrease in electron density after copper ion binding (**Figure 15**). In contrast, Zn^{2+} ion caused the enhancement in the fluorescence emission (~4.1-fold) of **36** due to the filled d^{10} electronic configuration of the Zn^{2+} ion, which does not usually involve energy or electron transfer mechanisms for the deactivation of the excited state (**Figure 15**).

Finally, the sensing mechanism of **36-Cu^{2+}** and **36-Zn^{2+}** has been shown to be reversible in the presence of EDTA, in which the fluorescence of **36** was almost recovered immediately from both complexes, which suggests the high reversibility of the chemosensor and the potential application in real-time monitoring [60].

4. Conclusion

As exposed in this chapter, hydrazone derivatives have been extensively employed as fluorescent and colorimetric chemosensors targeting important biological analytes such as inorganic anions and metal cations. Thus, it is clear that hydrazone scaffold is of great importance in the design of optical sensors. Here we have demonstrated just some representative examples of hydrazone and their ability as chemosensor for CN^-, F^-, AcO^-, multiple anions, Cu^{2+}, Zn^{2+}, Hg^{2+}, Al^{3+}, and multiple metals. Furthermore, we hope that this book chapter with discussions about the sensing mechanisms (PET, FRET, ESIPT, etc.) could be an important tool and contribute to the development of new rational research projects with the hydrazine scaffold for biological and environmental monitoring of metals and anions.

Acknowledgements

This work was supported by Fundação de Amparo à Pesquisa do Estado do Rio de Janeiro (FAPERJ), Coordenação de Aperfeiçoamento de Pessoal de Nível Superior (CAPES), and Conselho Nacional de Desenvolvimento Científico e Tecnológico (CNPq).

Conflict of interest

The authors declare no conflict of interest.

Author details

Thiago Moreira Pereira and Arthur Eugen Kümmerle*
Rural Federal University of Rio de Janeiro, Seropédica, Brazil

*Address all correspondence to: akummerle@hotmail.com

References

[1] Rollas S, Küçükgüzel SG. Biological activities of hydrazone derivatives. Molecules. 2007;**12**:1910-1939. DOI: 10.3390/12081910

[2] Lazny R, Nodzewska A. N,N-dialkylhydrazones in organic synthesis. From simple N,N-dimethylhydrazones to supported chiral auxiliaries. Chemical Reviews. 2010;**110**:1386-1434. DOI: 10.1021/cr900067y

[3] Lehn JM. From supramolecular chemistry towards constitutional dynamic chemistry and adaptive chemistry. Chemical Society Reviews. 2007;**36**:151-160. DOI: 10.1039/b616752g

[4] Areas ES, Bronsato BJS, Pereira TM, Guedes GP, Miranda FS, Kummerle AE, et al. Novel CoIII complexes containing fluorescent coumarin-N-acylhydrazone hybrid ligands: Synthesis, crystal structures, solution studies and DFT calculations. Spectrochimica Acta Part A: Molecular and Biomolecular Spectroscopy. 2017;**187**:130-142. DOI: 10.1016/j.saa.2017.06.031

[5] Pereira TM, Vitório F, Amaral RC, Zanoni KPS, Iha NYM, Kümmerle AE. Microwave-assisted synthesis and photophysical studies of novel fluorescent N-acylhydrazone and semicarbazone-7-OH-coumarin dyes. New Journal of Chemistry. 2016;**40**:8846-8854. DOI: 10.1039/c6nj01532h

[6] Su X, Aprahamian I. Hydrazone-based switches, metallo-assemblies and sensors. Chemical Society Reviews. 2014;**43**:1963-1981. DOI: 10.1039/c3cs60385g

[7] Serbutoviez C, Bosshard C, Knoepfle G, Wyss P, Pretre P, Guenter P, et al. Hydrazone derivatives, an efficient class of crystalline materials for nonlinear optics. Chemistry of Materials. 1995;**7**:1198-1206. DOI: 10.1021/cm00054a020

[8] Singh RB, Jain P, Singh RP. Hydrazones as analytical reagents: A review. Talanta. 1982;**29**:77-84. DOI: 10.1016/0039-9140(82)80024-6

[9] Lawrenceab MAW, Lorraine SC, Wilson KA, Kirk W. Review: Voltammetric properties and applications of hydrazones and azo moieties. Polyhedron. 2019;**173**:114111

[10] Wang L, Chen X, Cao D. A cyanide-selective colorimetric "naked-eye" and fluorescent chemosensor based on a diketopyrrolopyrrole–hydrazone conjugate and its use for the design of a molecular-scale logic device. RSC Advances. 2016;**6**:96676-96685. DOI: 10.1039/c6ra21669b

[11] Fan L, Qin J, Li T, Wang B, Yangn Z. A chromone Schiff-base as Al(III) selective fluorescent and colorimetric chemosensor. Journal of Luminescence. 2014;**155**:84-88. DOI: 10.1016/j.jlumin.2014.06.023

[12] Qin J, Yang Z, Fan L, Cheng X, Li T, Wang B. Design and synthesis of a chemosensor for the detection of Al^{3+} based on ESIPT. Analytical Methods. 2014;**6**:7343-7348. DOI: 10.1039/c4ay01330a

[13] Puangploy P, Smanmoo S, Surareungchai W. A new rhodamine derivative-based chemosensor for highly selective and sensitive determination of Cu^{2+}. Sensors and Actuators B: Chemical. 2014;**193**:679-686. DOI: 10.1016/j.snb.2013.12.037

[14] Thota S, Rodrigues DA, Pinheiro PSM, Lima LM, Fraga CAM, Barreiro EJ. N-Acylhydrazones as drugs. Bioorganic & Medicinal Chemistry Letters. 2018;**28**:2797-2806. DOI: 10.1016/j.bmcl.2018.07.015

[15] Narang R, Narasimhan B, Sharma S. A review on biological activities and chemical synthesis of hydrazide derivatives. Current Medicinal Chemistry. 2012;**19**:569-612. DOI: 10.2174/092986712798918789

[16] Busschaert N, Caltagirone C, Rossom WV, Gale PA. Applications of supramolecular anion recognition. Chemical Reviews. 2015;**115**:8038-8155. DOI: 10.1021/acs.chemrev.5b00099

[17] Horváth M, Cigáň M, Filo J, Jakusová K, Gáplovský M, Šándrik R, et al. Isatin pentafluorophenylhydrazones: Interesting conformational change during anion sensing. RSC Advances. 2016;**6**:109742-109750. DOI: 10.1039/c6ra22396f

[18] Ali R, Gupta RC, Dwivedi SK, Misra A. Excited state proton transfer (ESIPT) based molecular probe to sense F^- and CN^- anions through a fluorescence "turn-on" response. New Journal of Chemistry. 2018;**42**: 11746-11754. DOI: 10.1039/c8nj01435c

[19] Zhou Y, Zhang JF, Yoon J. Fluorescence and colorimetric chemosensors for fluoride-ion detection. Chemical Reviews. 2014;**114**:5511-5571. DOI: 10.1021/cr400352m

[20] Hudnall TW, Chiu CW, Gabbai FP. Fluoride ion recognition by chelating and cationic boranes. Accounts of Chemical Research. 2009;**42**:388-397. DOI: 10.1021/ar8001816

[21] Gu JA, Mania V, Huang ST. Design and synthesis of ultrasensitive off–on fluoride detecting fluorescence probe via autoinductive signal amplification. The Analyst. 2015;**140**:346-352. DOI: 10.1039/C4AN01723D

[22] Wang F, Wang L, Chen X, Yoon J. Recent progress in the development of fluorometric and colorimetric chemosensors for detection of cyanide ions. Chemical Society Reviews. 2014;**43**:4312-4324. DOI: 10.1039/C4CS00008K

[23] Sun Y, Liu Y, Guo W. Fluorescent and chromogenic probes bearing salicylaldehyde hydrazone functionality for cyanide detection in aqueous solution. Sensors & Actuators: B. Chemical. 2009;**143**:171-176. DOI: 10.1016/j.snb.2009.09.038

[24] Sun Y, Liu Y, Chen M, Guo W. A novel fluorescent and chromogenic probe for cyanide detection in water based on the nucleophilic addition of cyanide to imine group. Talanta. 2009;**80**:996-1000. DOI: 10.1016/j.talanta.2009.08.026

[25] Lin Q, Liu X, Wei TB, Zhang YM. Reaction-based ratiometric chemosensor for instant detection of cyanide in water with high selectivity and sensitivity. Chemistry, an Asian Journal. 2013;**8**:3015-3021. DOI: 10.1002/asia.201300791

[26] Hu J, Li J, Qi J, Sun Y. Selective colorimetric and "turn-on" fluorimetric detection of cyanide using an acylhydrazone sensor in aqueous media. New Journal of Chemistry. 2015;**39**:4041-4046. DOI: 10.1039/c5nj00089k

[27] Zhang W, Xu K, Yue L, Shao Z, Feng Y, Fang M. Two-dimensional carbazole-based derivatives as versatile chemosensors for colorimetric detection of cyanide and two-photon fluorescence imaging of viscosity *in vitro*. Dyes and Pigments. 2017;**137**:560-568. DOI: 10.1016/j.dyepig.2016.11.002

[28] Long C, Hu JH, Ni PW, Yin ZY, Fu QQ. A novel colorimetric and ratiometric fluorescent CN^- sensor based on rhodamine B hydrazone derivatives in aqueous media and its application in sprouting potatoes. New Journal of Chemistry. 2018;**42**:17056-17061. DOI: 10.1039/c8nj01612g

[29] Lin Z, Ou S, Duan C, Zhang B, Bai Z. Naked-eye detection of fluoride ion in water: A remarkably selective easy-to-prepare test paper. Chemical Communications. 2006:624-626. DOI: 10.1039/b514337c

[30] Han F, Bao Y, Yang Z, Fyles TM, Zhao J, Peng X, et al. Simple bisthiocarbonohydrazones as sensitive, selective, colorimetric, and switch-on fluorescent chemosensors for fluoride anions. Chemistry—A European Journal. 2007;**13**:2880-2892. DOI: 10.1002/chem.200600904

[31] Yang X, Gong X, Li Y, Liu Z, Gao B, Zhang G, et al. Diketopyrrolopyrrole-based chemosensors for selective recognition of fluoride ions. Tetrahedron. 2015;**71**:5069-5077. DOI: 10.1016/j.tet.2015.05.079

[32] Huang W, Chen Z, Lin H, Lin H. A novel thiourea–hydrazone-based switch-on fluorescent chemosensor for acetate. Journal of Luminescence. 2011;**131**:592-596. DOI: 10.1016/j.jlumin.2010.10.036

[33] Qiao YH, Lin H, Shao J, Lin HK. A highly selective naked-eye colorimetric sensor for acetate ion based on 1,10-phenanthroline-2,9-dicarboxyaldehyde-di-(p-substitutedphenyl-hydrazone). Spectrochimica Acta Part A. 2009;**72**: 378-381. DOI: 10.1016/j.saa.2008.10.007

[34] Shao J, Qiao Y, Lin H, Lin H. A C 3-symmetric colorimetric anion sensor bearing hydrazone groups as binding sites. Spectrochimica Acta Part A: Molecular and Biomolecular Spectroscopy. 2009;**71**:1736-1740. DOI: 10.1016/j.saa.2008.06.025

[35] Suganya S, Park JS, Velmathi S. Visual sensing of aqueous anions by C2-symmetric chemosensor and its application in real sample analysis. Sensors & Actuators: B. Chemical. 2014;**190**:679-684. DOI: 10.1016/j.snb.2013.09.010

[36] Saini N, Wannasiri C, Chanmungkalakul S, Prigyai N, Ervithayasuporn V, Kiatkamjornwong S. Furan/thiophene-based fluorescent hydrazones as fluoride and cyanide sensors. Journal of Photochemistry & Photobiology A: Chemistry. 2019;**385**:112038. DOI: 10.1016/j.jphotochem.2019.112038

[37] Que EL, Domaille DW, Chang CJ. Metals in neurobiology: Probing their chemistry and biology with molecular imaging. Chemical Reviews. 2008;**108**:1517-1549. DOI: 10.1021/cr078203u

[38] Zhang G, Li Y, Xu J, Zhang C, Shuang S, Dong C, et al. Glutathione-protected fluorescent gold nanoclusters for sensitive and selective detection of Cu^{2+}. Sensors and Actuators B: Chemical. 2013;**183**:583-588. DOI: 10.1016/j.snb.2013.04.023

[39] Sivaraman G, Iniya M, Anand T, Kotla NG, Sunnapu O, Singaravadivel S, et al. Chemically diverse small molecule fluorescent chemosensors for copper ion. Coordination Chemistry Reviews. 2018;**357**:50-104. DOI: 10.1016/j.ccr.2017.11.020

[40] Jung JH, Lee JH, Shinkai S. Functionalized magnetic nanoparticles as chemosensors and adsorbents for toxic metal ions in environmental and biological fields. Chemical Society Reviews. 2011;**40**:4464-4474. DOI: 10.1039/C1CS15051K

[41] Kumar M, Kumar R, Bhalla V, Sharma PR, Kaur T, Qurishi Y. Thiacalix[4]arene based fluorescent probe for sensing and imaging of Fe^{3+} ions. Dalton Transactions. 2012;**41**:408-412. DOI: 10.1039/c1dt11169h

[42] Huang L, Cheng J, Xie K, Xi P, Hou F, Li Z, et al. Cu^{2+}-selective fluorescent chemosensor based on coumarin and its application in bioimaging. Dalton Transactions.

2011;**40**:10815-10817. DOI: 10.1039/c1dt11123j

[43] Maity D, Karthigeyan D, Kundu TK, Govindaraju T. FRET-based rational strategy for ratiometric detection of Cu^{2+} and live cell imaging. Sensors & Actuators: B. Chemical. 2013;**176**:831-837. DOI: 10.1016/j.snb.2012.09.071

[44] Hu S, Song J, Zhao F, Meng X, Wu G. Highly sensitive and selective colorimetric naked-eye detection of Cu^{2+} in aqueous medium using a hydrazone chemosensor. Sensors & Actuators: B. Chemical. 2015;**215**:241-248. DOI: 10.1016/j.snb.2015.03.059

[45] Mukherjee S, Betal S. Sensing phenomena, extraction and recovery of Cu^{2+} followed by smart phone application using a luminescent pyrene based chemosensor. Journal of Luminescence. 2018;**204**:145-153. DOI: 10.1016/j.jlumin.2018.07.038

[46] Pereira TM, Franco DP, Vitorio F, Kümmerle AE. Coumarin compounds in medicinal chemistry: Some important examples from the last years. Current Topics in Medicinal Chemistry. 2018;**18**:124-148. DOI: 10.2174/1568026618666180329115523

[47] Xu Z, Liu X, Pan J, Spring DR. Coumarin-derived transformable fluorescent sensor for Zn^{2+}. Chemical Communications. 2012;**48**:4764-4766. DOI: 10.1039/C2CC30963G

[48] Peng X, Tang X, Qin W, Dou W, Guo Y, Zheng J, et al. Aroylhydrazone derivative as fluorescent sensor for highly selective recognition of Zn^{2+} ions: Syntheses, characterization, crystal structures and spectroscopic properties. Dalton Transactions. 2011;**40**:5271-5277. DOI: 10.1039/c0dt01590c

[49] An J, Yan M, Yang Z, Li T, Zhou Q. A turn-on fluorescent sensor for Zn(II) based on fluorescein-coumarin conjugate. Dyes and Pigments. 2013;**99**:1-5. DOI: 10.1016/j.dyepig.2013.04.018

[50] Datta BK, Thiyagarajan D, Samanta S, Ramesh A, Das G. A novel chemosensor with visible light excitability for sensing Zn^{2+} in physiological medium and in HeLa cells. Organic & Biomolecular Chemistry. 2014;**12**:4975-4982. DOI: 10.1039/c4ob00653d

[51] He G, Zhao Y, He C, Liu Y, Duan C. "Turn-on" fluorescent sensor for Hg^{2+} via displacement approach. Inorganic Chemistry. 2008;**47**:5169-5176. DOI: 10.1021/ic702494s

[52] Park S, Kim W, Swamy KMK, Lee HY, Jung JY, Kim G, et al. Rhodamine hydrazone derivatives bearing thiophene group as fluorescent chemosensors for Hg^{2+}. Dyes and Pigments. 2013;**99**:323-328. DOI: 10.1016/j.dyepig.2013.05.015

[53] Kai Y, Yang S, Gao X, Hu Y. Colorimetric and "turn-on" fluorescent for Hg^{2+} based on rhodamine-3,4-ethylenedioxythiophene derivative. Sensors & Actuators: B. Chemical. 2014;**202**:252-256. DOI: 10.1016/j.snb.2014.04.089

[54] Kumari C, Sain D, Kumar A, Nayek HP, Debnath S, Saha P, et al. A bis-hydrazone derivative of 2,5-furandicarboxaldehyde with perfect hetero-atomic cavity for selective sensing of Hg(II) and its intracellular detection in living HeLa S3 cell. Sensors & Actuators: B. Chemical. 2017;**243**:1181-1190. DOI: 10.1016/j.snb.2016.12.103

[55] Yue X, Wang Z, Li C, Yang Z. Naphthalene-derived Al^{3+}-selective fluorescent chemosensor based on PET and ESIPT in aqueous solution. Tetrahedron Letters. 2017;**58**:4532-4537. DOI: 10.1016/j.tetlet.2017.10.044

[56] Zhang K, Yang Z, Wang B, Sun SB, Li YD, Li T, et al. A highly selective chemosensor for Al^{3+} based on 2-oxo-quinoline-3-carbaldehyde Schiff-base.

Spectrochimica Acta Part A: Molecular and Biomolecular Spectroscopy. 2014;**124**:59-63. DOI: 10.1016/j.saa.2013.12.076

[57] Liao ZC, Yang ZY, Li Y, Wang BD, Zhou QX. A simple structure fluorescent chemosensor for high selectivity and sensitivity of aluminum ions. Dyes and Pigments. 2013;**97**:124-128. DOI: 10.1016/j.dyepig.2012.12.017

[58] Fan L, Li T, Wang B, Yang Z, Liu C. A colorimetric and turn-on fluorescent chemosensor for Al(III) based on a chromone Schiff-base. Spectrochimica Acta Part A: Molecular and Biomolecular Spectroscopy. 2014;**118**:760-764. DOI: 10.1016/j.saa.2013.09.062

[59] Wang L, Yan J, Qin W, Liu W, Wang R. A new rhodamine-based single molecule multianalyte (Cu^{2+}, Hg^{2+}) sensor and its application in the biological system. Dyes and Pigments. 2012;**92**:1083-1090. DOI: 10.1016/j.dyepig.2011.07.010

[60] Anbu S, Ravishankaran R, Silva MFCG, Karande AA, Pombeiro AJL. Differentially selective chemosensor with fluorescence off–on responses on Cu^{2+} and Zn^{2+} ions in aqueous media and applications in pyrophosphate sensing, live cell imaging, and cytotoxicity. Inorganic Chemistry. 2014;**53**:6655-6664. DOI: 10.1021/ic500313m

Section 5

From Bioinformatics to Computational Biology

Chapter 7

Systems Glycobiology: Past, Present, and Future

Songül Yaşar Yıldız

Abstract

Glycobiology is a glycan-based field of study that focuses on the structure, function, and biology of carbohydrates, and glycomics is a sub-study of the field of glycobiology that aims to define structure/function of glycans in living organisms. With the popularity of the glycobiology and glycomics, application of computational modeling expanded in the scientific area of glycobiology over the last decades. The recent availability of progressive Wet-Lab methods in the field of glycobiology and glycomics is promising for the impact of systems biology on the research area of the glycome, an emerging field that is termed "systems glycobiology." This chapter will summarize the up-to-date leading edge in the use of bioinformatics tools in the field of glycobiology. The chapter provides basic knowledge both for glycobiologists interested in the application of bioinformatics tools and scientists of computational biology interested in studying the glycome.

Keywords: glycan, glycobiology, glycome, systems biology, systems glycobiology

1. Introduction

Glycans are long chains of carbohydrate-based polymers composed of repeating units of monosaccharide monomers bound together by glycosidic linkages. Complex and diverse glycans appear to be ever-present macromolecules in all cells in nature, and essential to all biological systems. Glycans play physical, structural, and metabolic roles in living organisms [1]. In the last century, knowledge on the biochemistry and biology of nucleic acids and proteins rapidly increased. Nevertheless, it has been much more difficult to understand the biology of glycans, which are main component of the cell surface [2]. The biosynthesis mechanism of glycans is totally different from those of nucleic acids and proteins. Biological mechanism of glycans is complex, which makes analysis of them extremely difficult and limits our understanding of mechanisms responsible for biological functions of glycans [3]. After the genomics revolution and development of high-throughput technologies, scientific interests increased to understand the characterization, function, and interaction of other significant biomolecules (e.g., DNA transcripts, proteins, lipids, and glycans) for the cell. These interests resulted in emergence of other omic types such as transcriptomics, proteomics, metabolomics, lipidomics and glycomics [4]. From the perspective of evolutionary conservation, conservation decreased in the order genomics, transcriptomics, proteomics, metabolomics, lipidomics, and glycomics. On the other hand, reverse order is present for informational diversity of these fields of omics (**Figure 1**) [5].

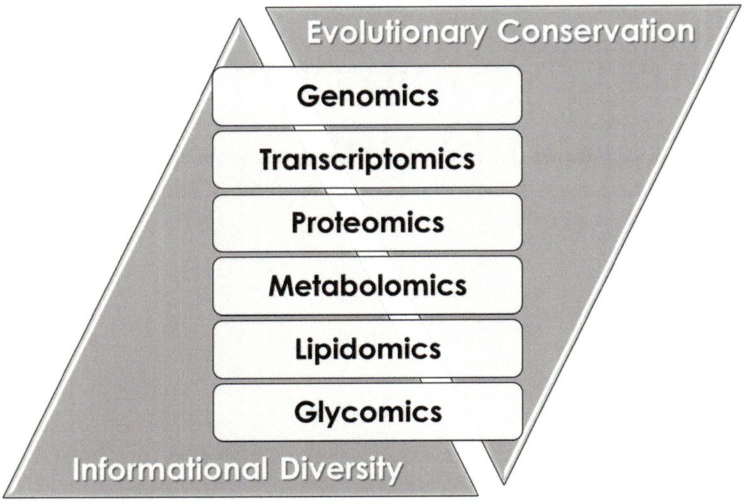

Figure 1.
The degree of evolutionary conservation and informational diversity for the omics fields.

With the progress in high-throughput technologies, studies on glycobiology increased to screen cells quickly and generate huge glycomics data sets. Moreover, advanced analytical techniques and tools for data analysis provide possibility to improve high-throughput techniques for screening glycans as a marker of diseases and to classify structure of glycans in therapeutic proteins [6].

2. Glycans

Glycans are linear or branched sugar macromolecules composed of repeating monosaccharides linked glycosidically. Beside nucleic acids and protein, glycans are known as the third dimension in molecular biology [7, 8]. These macromolecules can be found in the form of heteropolysaccharides or homopolysaccharides. Furthermore, glycoconjugates (glycolipid, glycoprotein and proteoglycan), can be also considered as glycan despite the fact that the carbohydrate part of glycoconjugates are only oligosaccharides [9]. In glycoproteins, oligosaccharides and proteins can be linked in different forms, namely N-linked glycans and O-linked glycans. N-acetylglucosamine is linked to the amide side chain of asparagine in N-linked glycans. C-1 of N-acetylgalactosamine is linked to the hydroxyl function of serine or threonine in O-linked glycans [10].

With the increasing researches in glycoscience, many different roles of glycans in biological systems have been revealed in the last decades. Significant functions of glycans have been determined in numerous research areas such as immunity, development and differentiation, biopharmaceuticals, cancer, fertilization, blood types, infectious diseases, etc. Glycans are called as "cloths of cells" since they are present on the surface of the cell and responsible for the signaling and communications between cells. Glycans can be classified in several ways. Varki divided the biological roles of glycans into four main categories: (1) structural and modulatory roles, (2) extrinsic (interspecies) recognition of glycans, (3) intrinsic (intraspecies) recognition of glycans, and (4) molecular mimicry of host glycans. A total of 50 distinct roles are defined under these main categories [1].

Glycans perform huge range of biological function due to the diversity of them, and they have significant roles in several physiological and pathological events, such as cell growth, cell signaling, cell-cell interactions, differentiation, and tumor growth [11–13]. In biological systems, information is carried by glycans, which are significant biomarker candidates for many diseases such as cardiovascular diseases, deficiencies of immune system, genetically inherited disorders, several cancer types, and neurodegenerative diseases [14–16]. Alteration of glycan expression is observed during the development and progression of these diseases, which is caused by misregulated enzymes such as glycosyltransferases and glycosidases. As a result, altered glycan structures have potential use for the identification of these diseases at an early stage. Besides significant role of glycans in diagnosis and management of disease, they can be used as therapeutics, markers for identification and isolation of special cell types, and targets in discovery of drugs [17–19]. Moreover, glycans can be considered as an ideal target for vaccines due to the presence of them on the surface of several different pathogens and malignant cells. High affinity and exquisite specificity of other molecules to recognize glycans are a vital point of developments in the research of glycans and related diagnostics and therapeutic applications.

3. Glycomics

Glycosylation plays significant roles in many biological processes including growth and development of cell, tumor growth and metastasis, immune recognition and response, intercommunication of cells, and microbial pathogenesis. As a result, glycosylation of proteins is the one of the most common and significant posttranslational modifications of proteins [20, 21]. Furthermore, more than half of proteins undergo glycosylation [6]. Many issues such as genetic factors, nucleotide levels of monosaccharides, cytokines, metabolites, hormones, and ecological factors can affect and change glycosylation process [20–24]. Thus, integration of omics approaches (e.g., proteomics, genomics, transcriptomics, and metabolomics) to the field of glycobiology is essential to view the big picture of the whole biological system [20, 21, 25]. Furthermore, for the analysis of glycans and glycosylation pathways, many glycoinformatics tools and databases are now accessible [6].

Glycomics is one of the most recent types of omics area which is responsible for the structure and function evaluations of glycans in bio-systems [26]. Integrating glycomics to other fields of omics provides new system-scale insights in integrative biology [27].

Moreover, glycomics informs other crucial scholarships such as systems glycobiology and personalized glycomedicine that collectively aim to explain the role of glycans in person-to-person and between population variations in disease susceptibility and response to health interventions such as drugs, nutrition, and vaccines. Glycosylation is present in both normal and diseased individuals [1]. Abnormal glycosylation is observed in a variety of diseases. Difference between glycosylation patterns of healthy and diseased individuals can be used as glycobiomarkers in personalized medicine [28]. As a result, many new medical implications will be enabled by glycobiology and glycopathology [29]. Development of glycomedicine can be contributed by holistic approach of functional and structural glycomics, which have applications in therapy development, fine-tuning immunological responses and the performance of therapeutic antibodies and boosting immune responses [28, 30]. Many applications of glycan arrays are present in many fields, from basic biochemical research to biomedical applications [31]. In addition to shotgun glycan microarrays [32], cell-based array resource has been developed [33]. These developments enable deeper understanding of the many biological roles of the glycome. Nevertheless, multiplatform and multiomics technologies are expected to further extend the knowledge of molecular mechanisms of glycans.

3.1 Major glycomics techniques

Monosaccharides represent four free hydroxyl groups for the linkage of another monosaccharide. As a result of this, glycans have more complex structure compared to structure of peptides and nucleic acids. It is known that glycans are more than the sequential monosaccharides; monomer types, modifications, the position of modifications around the ring of sugar, glycopolymer branching, and linkages chirality are the factors that are responsible for the complexity. As a result, sequencing techniques used for peptides or DNA (Sanger or Edman sequencing) are not appropriate for glycans. Moreover, most of the glycans are present as a part of a glycoconjugate. Therefore, glycan part should be released from lipid or protein part, by the use of enzymatic or chemical methods and isolated for analysis.

In the last decades, a number of techniques developed and applied to determine structure of the glycans with different degrees of detail [34]. A traditional method is to label the glycoconjugates radioactively and then apply anionic exchange, gel filtration, or paper chromatographic analyses prior and subsequent to enzymatic or chemical treatments. Still, it is difficult to figure out the definition of the actual structure; in consequence, in earlier studies, if adequate amounts were present, gas chromatography together with mass spectrometry (GC-MS) and/or nuclear magnetic resonance (NMR) studies were performed. However, these analyses involve special expertise to perform the research and interpret the results, particularly if standards were unavailable to compare with results.

HPLC and UPLC have superseded simple chromatography systems in recent years, and radioactive labeling has been replaced by fluorescent labeling. Nowadays, variable columns such as graphitized carbon, reversed-phase (RP), anion exchange, normal phase, or hydrophilic interaction resins can be used along with suitable enzymatic/chemical treatments. A less used alternative is to analyze glycans at elevated pH. As a result of this, the hydroxyl side chain deprotonation occurs, that enables the usage of anion exchange together with amperometric detection (HPAEC-PAD). On the other hand, glycan structure cannot be defined only by HPLC retention times, and for the unknown structure, analyses in the absence of standards should be interpreted with attention [35].

With the improvements in the types and the sensitivity, contribution of mass spectrometer to studies of glycans and glycoconjugates has increased in the last decades [36, 37]. At first, for the analysis of variable types of glycopolymers from different sources, researchers used fast atom bombardment mass spectrometers (FAB-MS). For the analyses with FAB-MS, chemical modifications such as methylation and acetylation were required. As an alternative method, matrix-assisted laser-desorption/ionization time-of-flight mass spectrometry (MALDI-TOF MS) was developed and analysis of both permethylated and native glycans can be performed with MALDI-TOF MS. Furthermore, current numerous electrospray techniques with many detector types have significance in glycomics. Mainly, a significant point in MS-based analysis is the capability to obtain glycan fragments. Besides, the preparation and separation techniques are of great importance to obtain the best results. As a consequence, liquid chromatography-mass spectrometry (LC-MS) in a number of forms is in general necessary since glycans with low abundancy or poor ionization capacity can be suppressed in the case of whole glycome examination. Moreover, reanalysis after the treatment of a chemical and an enzyme results in maximization of the ability to obtain clear results from the existing data.

Glycan is generally a part of the glycoconjugate; thus, glycoproteomics and glycolipidomics that consider both peptide/lipid and glycan parts are significant fields. At this point, mass spectrometry technique comes into prominence [38]. Both glycan and polypeptide/lipid parts can be studied with this technique. On the other hand, glycan parts of glycoproteins and glycolipids can be in various forms even if the polypeptide/lipid part is same, defined as microheterogeneity. The nature of glycan modifications is non-template driven and that leads to mentioned microheterogeneity [39].

Blotting technique can be used for simple screens. Reagents such as lectins and anti-carbohydrate antibodies with low specificity are often used for this technique; as a result, misleading results are often obtained [40]. Still, lectins, and antiserums have significance for immune responses in animals. New array-based systems can provide essential clues on proteins bounded to glycans [41].

4. Systems glycobiology and integration of omics data sets

Developments in integrative informatics and systems biology of glycans based on a holistic approach can make available a more comprehensive analysis. It elucidates annotation of glycans, enzyme levels, abundances of glycans biosynthesis pathways, and other omics data sets which are complementary. Though, several tools are developed for proteomics and genomics data sets and standard bioinformatics approaches are used in these tools, the complex relationships between diverse components (such as glycans, enzymes, transporters, and sugar nucleotides) of the glycosylation process are not considered by most of the existing bioinformatics tools. Consequently, the use of these tools for glycomics data sets has some limitations. The genome does not encode glycans directly and unlike proteins, interconnected action of many enzymes provides assembly of glycans. Due to mentioned limitations, developments in glycan analysis tools and methods have been delayed and most of the present glycoinformatics tools are special for single type data analysis [42–45]. For instance, database matching between obtained MS results and specific glycans in a glycan library is used as a mutual method for MS-based glycoprofiling for the purpose of individual peak annotation [46, 47]. If the complexity of glycosylation is wanted to be considered, enzymes of the organism which synthesize the studied glycans should generate glycan structures used for the annotation of the spectrum [48]. Due to this alignment, activities of enzyme and those structures assigned to each peak in the same spectrum will be consistent.

Although many omics approaches have significant progress in the last decades, existing techniques of bioinformatics are still unsatisfactory for the integration of varied data sets [49–51]. For instance, relations between expression levels of gene and specific glycan linkages abundance are investigated by statistical database-driven approaches, and these approaches could not predict quantity of detailed glycan distributions [50, 51]. This indicates the necessity of glycoinformatics and systems biology tools integration for the identification of glycan structure and these should be also linked to the information of gene expression responsible for glycosylation enzymes which synthesize these glycans. In order to understand levels of mRNA which is related with the distribution and quantity of glycans present within healthy and diseased cells, mathematical modeling of glycosylation is considered as a promising method [48, 49, 52].

Variability in the platform of analytical high-throughput experiments can be reduced by data integration approach. Increased confidence of biomarker predictions and recommendations can be obtained if different data from experiments such as glycogenes expression information or mass spectra profile confirm the results from integrative glycoinformatics and systems glycobiology tools. Although integrated glycoinformatics tools have limitations in analytical sensitivities, analysis and comparison of various results with various platforms are enabled by these tools [6].

The integration of glycomics with other various omics data is promising for further innovation in diagnosis and treatment of diseases [30]. The start point of multiomics data integration is to sort the data based on the omics level. In the following part, association between glycomics and other omics levels will be represented.

4.1 Genomics

Integration of glycomics with genetic sequence can be occur in a number of ways. For instance, glycosylation site can be gained or lost with the variation of sequence. A single-nucleotide polymorphism (SNP) affects glycosylation of prostate-specific antigen (PSA) and an altered function of it increases the risk of prostate cancer. Functional analysis indicated that the stability and structural conformation of PSA are affected by missense variant rs61752561, which causes an additional extra glycosylation site [53]. Furthermore, computational studies revealed that variations in cancer somatic cells have potential to cause gain or loss of glycosylation. In addition to SNP, variations in structure and abnormalities in cytogenetics could be integrated with glycomics. Cytogenetic abnormalities have been associated with glycome expression [54]. A particular glycosyltransferase can glycosylate numerous proteins, so genetic variants of it have extraordinary significance because function of many glycoproteins can be affected by a single difference in activity of enzyme. Several downstream pathways and cell metabolism can be affected by a genetic or epigenetic variant that is called pleiotropic effect of genetic or epigenetic variant on glycosylation [55].

4.2 Transcriptomics

Most of the glycomics research have been done at the level of transcriptome, which can be performed either at a particular locus or with a technology of microarray. In colorectal cancer (CRC), glycosyltransferase ST6GAL1 is associated with cancer, and altered ST6GAL1 expression was found by The Cancer Genome Atlas (TCGA) mining [56]. Moreover, in order to identify differential expression of glycosylation-related genes Saravanan et al. [57] used GLYCOv2 glycogene microarray technology. In the further studies, myeloma was compared with normal plasma cell samples and 60 upregulated and 20 downregulated genes were found among 243 genes in glycan-biosynthesis pathway [54]. A novel molecular signature that is enriched for enzymes of glycosylation was revealed by meta-analysis performed for gene expression of prostate cancer [58]. Additionally, hepatocellular carcinoma was investigated by reviewing gene expressions that are related with core fucosylation of the disease [59]. More systematic reviews and meta-analyses are required to develop reliable biomarkers.

4.3 Glycoproteomics

Studies on glycoproteomics include peptide structures, glycan structures, and sites of glycosylation [30]. Single site on the peptide chain can be glycosylated by different glycans, and by this way, glycans can modulate function of the protein [60]. In the literature, diverse techniques were associated with different phenotypes, for instance, breast cancer, colon cancer, liver cancer, skin cancer, ovary cancer, bladder cancer, and neurodegenerative diseases, and additionally, a number of structural variations including sialylation, fucosylation, degree of branching, and specific glycosyltransferases expression [61–63]. For instance, cerebrospinal fluid N-glycoproteomics is of significant importance in early diagnosis of Alzheimer's disease. Glycosylation patterns were assessed in patients and therapeutics targets such as glycoenzymes were suggested [63]. For the diagnosis of pancreatic cancer, specific glycoforms together with protein levels should be measured to improve potential for diagnosis [64]. Glycoproteins constitute the majority of protein tumor markers approved by Food and Drug Administration (FDA), and they are also used currently in clinical practice. Many of these glycoproteins have alterations of glycosylation in cancer [60]. MUC-1 (CA15-3/CA27.29) [65] and plasminogen activator inhibitor (PAI-1) [66] are biomarkers of breast cancer; beta-human chorionic gonadotropin (Beta-hCG) [67] is biomarker of colorectal cancer; alpha-fetoprotein (AFP) [68] is a biomarker of liver cancer and germ cell tumors; chromogranin A (CgA) [69] is a biomarker of neuroendocrine tumors; MUC16 (CA-125) [70] and HE4 [71] are biomarkers of ovarian cancer; and many other biomarkers are present for a variety type of cancer. Most of the results in the existing publications are heterogeneous; thus, systematic integrative reviews of the literature are required for further development of glycoproteomics.

4.4 Metabolomics

Metabolomics is the large-scale study of the small molecule substrates that investigates variations in the metabolites within cells, biofluids, tissues, or organism. Metabolomics and glycomics were investigated in the research of post-traumatic stress [72]. According to the researchers of this study, these biomarkers together with omics markers should be integrated to understand the biological differences responsible for this stress. For discovery of liver cancer biomarker, proteomics, glycomics, and metabolomics were integrated and this integration enhanced performance when compared to separate omics data [73]. Physiological and pathological conditions are reflected by metabolomic and glycomic data in individuals. Similar to metabolites, small glycans can be quantified easily [74]. Human Metabolome Database (HMDB) is the most inclusive metabolite source that offers significant resource for the discovery of biomarkers in glycomics [75].

4.5 Glycolipidomics

Glycolipidomics is a scientific field that identifies and quantifies glycolipids. For the determination of physiological and pathological conditions of individual, glycolipids can be used as a specific biomarker. They take role in development of neurological and neurodegenerative diseases, such as Lewy body dementia, Alzheimer's disease, Parkinson's disease, and frontotemporal dementia [30]. Furthermore, glycosphingolipids are associated with cancer and they are promising molecules for diagnosis as biomarkers and for malignant tumor immunotherapy as target [76]. More recently, Dehelean et al. [77] reviewed trends in the discovery of glycolipid biomarker by MS.

4.6 Interactomics

Interactomics is the research field that investigates whole set of interactions between molecules including glycans. Interaction of glycans with glycan-binding proteins (GBPs) is of significant importance in immune response, signaling, cell recognition, infections, neurodegenerative diseases, and cancer. High-throughput technologies ease studies also on interactomics [78]. UniLectin3D is a database that catalog lectins that are most studied GBPs. Database consists of curated information on 3D structures and interacting ligands [79]. Lectin-glycan interaction on surface of the cell is a significant factor for the regulation in corneal biology (i.e., corneal infection) and pathophysiology (i.e., inflammation) [80]. The whole protein-glycan interactome information has not been obtained yet [41]. For future studies, estimated number of interactions is of importance. GenProBiS is a bioinformatics tool that analyzes binding sites between peptide-peptide, peptide-nucleic acid, and peptide-compound and also sites of glycosylation and other posttranslational modifications. Furthermore, it provides maps between sequence variations and structure of protein. More developments of bioinformatics tools analyzing huge data will prioritize the objections for experimental verification and provide contribution to interactomics development.

4.7 Other omics fields

In future studies, many other omics fields should be associated with glycomics such as comparative genomics, epigenomics, regulomics, NcRNomics, MiRNomics, LncRNomics, etc. Although glycomics is the significant field related with molecular interactions, information about how these complex processes controlled by regulatory network is still inadequate. In addition to classic omics fields, omics applications such as iatromics, environmental omics, pharmacogenomics, and nutrigenomics should also be reviewed.

5. Bioinformatics tools and databases

Glycoinformatics combines bioinformatics tools with glycome. Glycomics data is collected by the tools and databases to investigate, reveal, and associate with other repository of related data of proteomics, genomics, and interactomics. Commonly used tools and databases are summarized in **Table 1**.

6. Current bottlenecks for systems glycobiology

System-based analyses applied smoothly to network of signaling, metabolic processes, and physiological modeling; however, applications in systems glycobiology still have problems in computational and analytical studies and this situation arises from prominent bottlenecks [81]: (i) there is no accepted standard for model building; (ii) glycoinformatics databases are underdeveloped; (iii) and insufficient quantitative data are from glycoproteomics experiments.

In recent years, many systems based models have been developed to simulate biosynthesis of glycans. Nevertheless, difficulty in the incorporation of glycan structure and specificity data of enzymes related with glycosylation into mathematical models. As a result of this difficulty, systematic model building is still not present in this field. Moreover, limited number of the current models is available in Systems Biology Markup Language (SBML) format [82], which is the obstacle to develop, share, and validate computational models.

In the last decades, many databases related to glycoscience have emerged. Nevertheless, functional information is limited when compared to glycan structure and taxonomy data. In the future, relation of glycan structure to specific enzymes that synthesize them, the rates of their synthesis, and also their function are required in order to build model.

	Name	Description	Link
Databases	CAZY	Describes the families of structurally related catalytic and carbohydrate binding modules (or functional domains) of enzymes that degrade, modify, or create glycosidic bonds	http://www.cazy.org/
	KEGG GLYCAN	The KEGG GLYCAN structure database is a collection of experimentally determined glycan structures. It contains all unique structures taken from CarbBank, structures entered from recent publications, and structures present in KEGG	https://www.genome.jp/kegg/glycan/
	Glycan Library	A list of approximately 830 lipid-linked sequence-defined glycan probes derived from diverse natural sources or chemically synthesized	https://glycosciences.med.ic.ac.uk/glycanLibraryIndex.html
	GlycoMob	An ion mobility-mass spectrometry collision cross-section database for glycomics	http://www.glycomob.org
	GlycoBase 3.2	A database of over 650 N- and O-linked glycan structures of HPLC, UPLC, exoglycosidase sequencing, and mass spectrometry (MALDI-MS, ESI-MS, ESI-MS/MS, LC-MS, LC-ESI-MS/MS) data	https://glycobase.nibrt.ie/glycobase/show_nibrt.action
	Glyco3D	A portal of 3D structures of mono-, di-, oligo-, and polysaccharides and carbohydrate recognizing proteins (lectins, monoclonal antibodies, glycosyltransferases) and glycosaminoglycan binding proteins	http://glyco3d.cermav.cnrs.fr/home.php
	GlyMAP	An online resource mapping of the variational landscape of glycoactive enzymes	http://glymap.glycomics.ku.dk/
	Glycosciences.de	A collection of databases and bioinformatics tools for glycobiology and glycomics	http://glycosciences.de/index.php
	UniProtKB	The universal protein sequence database with information on glycosylated proteins	http://www.uniprot.org/
	UniCarbKB	UniCarbKB is a curated and annotated glycan database which curates information from the scientific literature on glycoprotein-derived glycan structures. It includes data previously available from GlycoSuiteDB	http://www.unicarbkb.org/
	UniCarbDB	UniCarbDB is a platform for presenting glycan structures and fragment data characterized by LC-MS/MS strategies. The database is annotated with high-quality datasets and is designed to extend and reinforce those standards and	http://unicarb-db.biomedicine.gu.se/

Name	Description	Link
	ontologies developed by existing glycomics databases	
UniPep	A database for human N-linked glycosites: a resource for biomarker discovery	http://www.unipep.org
SugarBindDB	SugarBindDB provides a collection of known carbohydrate sequences to which pathogenic organisms specifically adhere via lectins or adhesins. The data were compiled through an exhaustive search of literature published over the past 30 years by glycobiologists, microbiologists, and medical histologists	http://sugarbind.expasy.org/
Consortium for Functional Glycomics (CFG)	The CFG serves to combine the expertise and glycomics resources to reveal functions of glycans and glycan-binding proteins (GBPs) that impact human health and disease. The CFG offers resources to the community, including glycan array screening services, a reagent bank, and access to a large glycomics database and data analysis tools	http://www.functionalglycomics.org/
GLYCONAVI	A Website for carbohydrate research. It consists of the "GlycoNAVI database" for molecular information of carbohydrates, and chemical reactions of carbohydrate synthesis, the "Route Searching System for Glycan Synthesis," and "GlycoNAVI tools" for editing two-dimensional molecular structure of carbohydrates	http://www.glyconavi.org/GlycoNAVI
GlycoGeneDataBase (GGDB)	Glycogene includes genes associated with glycan synthesis such as glycosyltransferase, sugar nucleotide synthases, sugar-nucleotide transporters, sulfotransferases, etc.	https://acgg.asia/ggdb2
Carbohydrate Structure Data Base (CSDB)	CSDB covers information on structures and taxonomy of natural carbohydrates published in the literature and mostly resolved by nuclear magnetic resonance (NMR). CSDB is composed of two parts: Bacterial and Archeal (BCSDB) and Plant and Fungal (PFCSDB)	http://csdb.glycoscience.ru/database/core/help.php?topic=rules
EXPASy	This section of the ExPASy server gathers a toolbox for processing data as well as simulating, predicting, or visualizing information, relative to glycans, glycoproteins, and glycan-binding proteins	http://www.expasy.org/glycomics

	Name	Description	Link
TOOLS	CASPER	A tool for calculating NMR chemical shifts of oligo- and polysaccharides	http://www.casper.organ.su.se/casper/
	Glycan Builder	A software library and set of tools to allow the rapid drawing of glycan structures with support for all of the most common symbolic notation formats	http://www.unicarbkb.org/builder
	GlycoDomainViewer	An online resource to study site glycosylation with respect to protein context and conservation	http://glycodomain.glycomics.ku.dk/
	Glynsight	Glynsight offers visualization and interactive comparison of glycan expression profiles. The tool was initially developed with a focus on IgG N-glycan profiles but it was extended to usage with any experiment, which produces N- or O-linked glycan expression data	https://glycoproteome.expasy.org/glynsight/
	GlycoMinestruct	A new bioinformatics tool for highly accurate mapping of the human N-linked and O-linked glycoproteomes by incorporating structural features	http://glycomine.erc.monash.edu/Lab/GlycoMine_Struct/
	GlyMAP	An online resource mapping out the variational landscape of glycoactive enzymes	http://glymap.glycomics.ku.dk/
	GlycoMod	An online tool to predict oligosaccharide structures on proteins from experimentally determined masses	http://web.expasy.org/glycomod/
	GlycoMiner/GlycoPattern	Software tools designed to detect, characterize, and perform relative quantitation of N-glycopeptides based on LC-MS runs	http://www.szki.ttk.mta.hu/ms/glycominer/
	Glycosciences.de	A collection of databases and bioinformatics tools for glycobiology and glycomics	http://glycosciences.de/index.php
	RINGS	A Web resource providing algorithmic and data mining tools to aid glycobiology research	http://rings.t.soka.ac.jp/
	MonosaccharideDB	A comprehensive reference source for monosaccharide notation	http://www.monosaccharidedb.org/start.action
	NetOGlyc	Next generation prediction of O-glycosylation sites on proteins	http://www.cbs.dtu.dk/services/NetOGlyc/
	GlycoSpectrumScan	A Web-based bioinformatic tool designed to link glycomics and proteomics analyses for the characterization of glycopeptides. GlycoSpectrumScan is a MS platform which is independent, freely accessible, and profiles glycopeptide MS data using beforehand separately acquired released glycan and proteomics	https://github.com/wliu1197/glycospectrumscan

Name	Description	Link
	information. Both N- and O-glycosylated peptides as well as multiply glycosylated peptides can be analyzed	
SimGlycan	A predictive carbohydrate analysis tool for MS/MS data	http://www.premierbiosoft.com/glycan
SugarQb	SugarQb enables genome-wide insights into protein glycosylation and glycan modifications in complex biological systems. This is a collection of software tools (Nodes) which enable the automated identification of intact glycopeptides from HCD-MS/MS data sets, using commonly use peptide-centric MS/MS search engines	http://www.imba.oeaw.ac.at/SugarQb
GlycoDigest	GlycoDigest is a tool that simulates exoglycosidase digestion based on controlled rules acquired from expert knowledge and experimental evidence available in GlycoBase	www.glycodigest.org/
Virtual Glycome	This Website is focused on presenting selected computational tools and experimental resources that can be used to better understand the processes regulating cellular glycosylation at multiple levels	https://virtualglycome.org/
SweetUnityMol	Software to display 3-D structures of carbohydrates, polysaccharides, and glycoconjugates	http://sourceforge.net/projects/unitymol/files/

Table 1.
Tools and databases used in glycoinformatics.

For the measurement of glycome, two main approaches are common. In the first approach, enzymes or mild hydrolysis is used to separate the glycans from the peptide backbone. Next, to obtain information about the composition and relative abundance of the carbohydrate structures, permethylation of glycans and MS analysis are used [83]. The bottleneck is the lack of well-developed software. For the data analysis of glycoproteomics and correspondingly acceleration of system-based model building and validation, more sophisticated computational tools are required.

7. Mathematical modeling of biochemical reaction networks

Mathematical models of glycosylation are developed in three main stages: (i) biological information gathering; (ii) model formulation; (iii) and simulation and postsimulation analysis. First step includes definition of components (enzyme, substrate, and product) crucial for the model. All of the components present in the biochemical network and connections between them are cataloged in this step. The process relies on information of biochemistry and cell biology, and analytical tools. In the next step, behavior in the steady state of the system is investigated by using simple linear algebra and principles of optimization. If time is a variable, the computer model can incorporate ordinary differential equations (ODEs) or Boolean networks. Proper kinetic/thermodynamic/stochastic/optimization parameters are collated depending on the formulation nature of the model and processes which are specified by enzymatically/nonenzymatically. The last step is performed to simulate the experimental system in the computer and to define unknown parameters of model by the help of fitting experimental data [81]. Visualization of multidimensional results is significant because numerous diverse models may attempt to fit one data set obtained from time labor and concentration-dependent experiments. As a result, consolidation of the findings obtained by simulations of complex reaction network and generation of hypotheses that can be tested experimentally require network analysis strategies.

8. Conclusions

Glycomics is a very comprehensive research area of science and interacts with several different omics fields. As many other omics types, it consists of a huge number of genomics components. In the future, techniques in high-throughput analysis and bioinformatics will be developed and enable the integration of all available data of glycomics into a particular diagram and by this way, it will be possible to develop biomarker and identify potential new therapeutic targets. Moreover, progresses in the field reveal that integrative multiomics approach should include glycomics in order to develop new biomarkers for robust diseases. One of the specific fields of systems biology is the systems glycobiology. It is based on a holistic approach that indicates process of complex glycosylation and associations between its constituents. A more complete glycome overview is targeted by using enzyme levels, abundances of glycans, pathways for biosynthesis, glycan annotation, and related omics data sets.

An approach of systems glycobiology is constructed in combination of various data sets of glycomics with that of other omics fields by the use of glycoinformatics tools to clarify understanding on process of glycosylation from various data sets. With the presented chapter, main aspects of glycobiology, glycomics, and systems glycobiology are summarized. However, these fields are still developing and further developments provide more insight to this specific research area.

Author details

Songül Yaşar Yıldız
Faculty of Engineering and Natural Sciences, Department of Bioengineering, İstanbul Medeniyet University, İstanbul, Turkey

*Address all correspondence to: songul.yildiz@medeniyet.edu.tr

References

[1] Varki A. Biological roles of glycans. Glycobiology. 2017;**27**(1):3-491

[2] Ohtsubo K, Marth JD. Glycosylation in cellular mechanisms of health and disease. Cell. 2006;**126**(5):855-867

[3] York WS, Kochut KJ, Miller JA. Integration of Glycomics Knowledge and Data. Handbook of Glycomics. Amsterdam, The Netherlands: Elsevier; 2010. pp. 177-195

[4] Ferreira CR, Turco L, Guimarães E, Saraiva SA, Bertolla RP, Perecin F, et al. Proteomics, metabolomics and lipidomics in reproductive biotechnologies: The MS solutions. Acta Scientiae Veterinariae. 2010;**38**:s591-s603

[5] Varki A. Evolutionary forces shaping the Golgi glycosylation machinery: Why cell surface glycans are universal to living cells. Cold Spring Harbor Perspectives in Biology. 2011;**3**(6):a005462

[6] Bennun SV, Hizal DB, Heffner K, Can O, Zhang H, Betenbaugh MJ. Systems glycobiology: Integrating glycogenomics, glycoproteomics, glycomics, and other 'omics data sets to characterize cellular glycosylation processes. Journal of Molecular Biology. 2016;**428**(16):3337-3352

[7] Yildiz SY, Erginer M, Demirci T, Hemberger J, Oner ET. Glycan-Based Nanocarriers in Drug Delivery. Drug Delivery Approaches and Nanosystems. Vol. 2. Florida (USA): Apple Academic Press; 2017. pp. 167-203

[8] Panitch A, Paderi JE, Sharma S, Stuart KA, Vazquez-Portalatin NM. Extracellular Matrix-Binding Synthetic Peptidoglycans. IN (US): Google Patents; 2018

[9] Dwek RA. Glycobiology: Toward understanding the function of sugars. Chemical Reviews. 1996;**96**(2):683-720

[10] Gorelik E, Galili U, Raz A. On the role of cell surface carbohydrates and their binding proteins (lectins) in tumor metastasis. Cancer and Metastasis Reviews. 2001;**20**(3–4):245-277

[11] Tommasone S, Allabush F, Tagger YK, Norman J, Köpf M, Tucker JH, et al. The challenges of glycan recognition with natural and artificial receptors. Chemical Society Reviews. 2019;**48**(22):5488-5505

[12] Lau KS, Partridge EA, Grigorian A, Silvescu CI, Reinhold VN, Demetriou M, et al. Complex N-glycan number and degree of branching cooperate to regulate cell proliferation and differentiation. Cell. 2007;**129**(1): 123-134

[13] Tian Y, Zhang H. Glycoproteomics and clinical applications. Proteomics – Clinical Applications. 2010;**4**(2):124-132

[14] Hwang H, Zhang J, Chung KA, Leverenz JB, Zabetian CP, Peskind ER, et al. Glycoproteomics in neurodegenerative diseases. Mass Spectrometry Reviews. 2010;**29**(1): 79-125

[15] Lowe JB, Marth JD. A genetic approach to mammalian glycan function. Annual Review of Biochemistry. 2003;**72**(1):643-691

[16] Adamczyk B, Tharmalingam T, Rudd PM. Glycans as cancer biomarkers. Biochimica et Biophysica Acta (BBA) - General Subjects. 2012; **1820**(9):1347-1353

[17] Hudak JE, Bertozzi CR. Glycotherapy: New advances inspire a reemergence of glycans in medicine. Chemistry & Biology. 2014;**21**(1):16-37

[18] Lanctot PM, Gage FH, Varki AP. The glycans of stem cells. Current

Opinion in Chemical Biology. 2007;**11**: 373-380

[19] Vasconcelos-dos-Santos A, Oliveira IA, Lucena MC, Mantuano NR, Whelan SA, Dias WB, et al. Biosynthetic machinery involved in aberrant glycosylation: Promising targets for developing of drugs against cancer. Frontiers in Oncology. 2015;**5**:138

[20] Raman R, Raguram S, Venkataraman G, Paulson JC, Sasisekharan R. Glycomics: An integrated systems approach to structure-function relationships of glycans. Nature Methods. 2005; **2**(11):817

[21] Liu L, Telford JE, Knezevic A, Rudd PM. High-Throughput Glycoanalytical Technology for Systems Glycobiology. London, UK: Portland Press Limited; 2010

[22] Butler M, Quelhas D, Critchley AJ, Carchon H, Hebestreit HF, Hibbert RG, et al. Detailed glycan analysis of serum glycoproteins of patients with congenital disorders of glycosylation indicates the specific defective glycan processing step and provides an insight into pathogenesis. Glycobiology. 2003; **13**(9):601-622

[23] Lauc G, Rudan I, Campbell H, Rudd PM. Complex genetic regulation of protein glycosylation. Molecular BioSystems. 2010;**6**(2):329-335

[24] Soo EC, Hui JP. Metabolomics in glycomics. In: Functional Glycomics. Berlin, Germany: Springer; 2010. pp. 175-186

[25] Zhang W, Li F, Nie L. Integrating multiple 'omics' analysis for microbial biology: Application and methodologies. Microbiology. 2010;**156**(2):287-301

[26] Adua E, Russell A, Roberts P, Wang Y, Song M, Wang W. Innovation analysis on postgenomic biomarkers: Glycomics for chronic diseases. OMICS: A Journal of Integrative Biology. 2017; **21**(4):183-196

[27] Ly M, Laremore TN, Linhardt RJ. Proteoglycomics: Recent progress and future challenges. OMICS: A Journal of Integrative Biology. 2010;**14**(4):389-399

[28] Reily C, Stewart TJ, Renfrow MB, Novak J. Glycosylation in health and disease. Nature Reviews Nephrology. 2019;**1**

[29] Gabius H-J, Kayser K. Introduction to glycopathology: The concept, the tools and the perspectives. Diagnostic Pathology. 2014;**9**(1):4

[30] Kunej T. Rise of systems Glycobiology and personalized Glycomedicine: Why and how to integrate Glycomics with multiomics science? OMICS. 2019;**23**(12):615-622

[31] Geissner A, Seeberger PH. Glycan arrays: From basic biochemical research to bioanalytical and biomedical applications. Annual Review of Analytical Chemistry. 2016;**9**: 223-247

[32] Smith DF, Cummings RD, Song X. History and future of shotgun glycomics. Biochemical Society Transactions. 2019;**47**(1):1-11

[33] Narimatsu Y, Joshi HJ, Nason R, Van Coillie J, Karlsson R, Sun L, et al. An atlas of human glycosylation pathways enables display of the human glycome by gene engineered cells. Molecular Cell. 2019;**75**(2):394-407. e5

[34] Geyer H, Geyer R. Strategies for analysis of glycoprotein glycosylation. Biochimica et Biophysica Acta (BBA)-Proteins and Proteomics. 2006; **1764**(12):1853-1869

[35] Wilson IB. Molecular parasitology. In: Glycomics. Berlin, Germany: Springer; 2016. pp. 75-89

[36] Alley WR Jr, Novotny MV. Structural glycomic analyses at high sensitivity: A decade of progress. Annual Review of Analytical Chemistry. 2013;**6**:237-265

[37] Haslam SM, Morris HR, Dell A. Mass spectrometric strategies: Providing structural clues for helminth glycoproteins. Trends in Parasitology. 2001;**17**(5):231-235

[38] Thaysen-Andersen M, Packer NH. Advances in LC–MS/MS-based glycoproteomics: Getting closer to system-wide site-specific mapping of the N-and O-glycoproteome. Biochimica et Biophysica Acta (BBA)-Proteins and Proteomics. 2014;**1844**(9):1437-1452

[39] Schachter H. Biosynthetic controls that determine the branching and microheterogeneity of protein-bound oligosaccharides. Biochemistry and Cell Biology. 1986;**64**(3):163-181

[40] Iskratsch T, Braun A, Paschinger K, Wilson IB. Specificity analysis of lectins and antibodies using remodeled glycoproteins. Analytical Biochemistry. 2009;**386**(2):133-146

[41] Cummings RD, Pierce JM. The challenge and promise of glycomics. Chemistry & Biology. 2014;**21**(1):1-15

[42] Ceroni A, Maass K, Geyer H, Geyer R, Dell A, Haslam SM. GlycoWorkbench: A tool for the computer-assisted annotation of mass spectra of glycans. Journal of Proteome Research. 2008;**7**(4):1650-1659

[43] Maass K, Ranzinger R, Geyer H, von der Lieth CW, Geyer R. "Glyco-peakfinder"–De novo composition analysis of glycoconjugates. Proteomics. 2007;**7**(24):4435-4444

[44] Goldberg D, Sutton-Smith M, Paulson J, Dell A. Automatic annotation of matrix-assisted laser desorption/ionization N-glycan spectra. Proteomics. 2005;**5**(4):865-875

[45] Kawano S, Hashimoto K, Miyama T, Goto S, Kanehisa M, editors. Prediction of glycan structures from DNA microarray data. In: Glycobiology. NC, USA: Journals Department, Oxford University Press; 2004

[46] An HJ, Lebrilla CB. A glycomics approach to the discovery of potential cancer biomarkers. In: Functional Glycomics. Berlin, Germany: Springer; 2010. pp. 199-213

[47] Joshi HJ, Harrison MJ, Schulz BL, Cooper CA, Packer NH, Karlsson NG. Development of a mass fingerprinting tool for automated interpretation of oligosaccharide fragmentation data. Proteomics. 2004;**4**(6):1650-1664

[48] Krambeck FJ, Betenbaugh MJ. A mathematical model of N-linked glycosylation. Biotechnology and Bioengineering. 2005;**92**(6):711-728

[49] Bennun SV, Yarema KJ, Betenbaugh MJ, Krambeck FJ. Integration of the transcriptome and glycome for identification of glycan cell signatures. PLoS Computational Biology. 2013;**9**(1): e1002813

[50] Kawano S, Hashimoto K, Miyama T, Goto S, Kanehisa M. Prediction of glycan structures from gene expression data based on glycosyltransferase reactions. Bioinformatics. 2005;**21**(21):3976-3982

[51] Suga A, Yamanishi Y, Hashimoto K, Goto S, Kanehisa M. An improved scoring scheme for predicting glycan structures from gene expression data. Genome Informatics. 2007;**18**:237-246

[52] Krambeck FJ, Bennun SV, Narang S, Choi S, Yarema KJ, Betenbaugh MJ. A mathematical model to derive N-glycan structures and cellular enzyme activities from mass spectrometric data. Glycobiology. 2009;**19**(11):1163-1175

[53] Srinivasan S, Stephens C, Wilson E, Panchadsaram J, DeVoss K,

Koistinen H, et al. Prostate cancer risk-associated single-nucleotide polymorphism affects prostate-specific antigen glycosylation and its function. Clinical Chemistry. 2019;**65**(1):e1-e9

[54] Moehler TM, Seckinger A, Hose D, Andrulis M, Moreaux J, Hielscher T, et al. The glycome of normal and malignant plasma cells. PLoS One. 2013; **8**(12):e83719

[55] Vojta A, Samaržija I, Bočkor L, Zoldoš V. Glyco-genes change expression in cancer through aberrant methylation. Biochimica et Biophysica Acta (BBA) - General Subjects. 2016; **1860**(8):1776-1785

[56] Venturi G, Gomes Ferreira I, Pucci M, Ferracin M, Malagolini N, Chiricolo M, et al. Impact of sialyltransferase ST6GAL1 overexpression on different colon cancer cell types. Glycobiology. 2019; **29**(10):684-695

[57] Saravanan C, Cao Z, Head SR, Panjwani N. Analysis of differential expression of glycosyltransferases in healing corneas by glycogene microarrays. Glycobiology. 2010;**20**(1): 13-23

[58] Barfeld SJ, East P, Zuber V, Mills IG. Meta-analysis of prostate cancer gene expression data identifies a novel discriminatory signature enriched for glycosylating enzymes. BMC Medical Genomics. 2014;**7**(1):513

[59] Norton PA, Mehta AS. Expression of genes that control core fucosylation in hepatocellular carcinoma: Systematic review. World Journal of Gastroenterology. 2019;**25**(23):2947

[60] Lauc G, Pezer M, Rudan I, Campbell H. Mechanisms of disease: The human N-glycome. Biochimica et Biophysica Acta (BBA) - General Subjects. 2016;**1860**(8):1574-1582

[61] Azevedo R, Peixoto A, Gaiteiro C, Fernandes E, Neves M, Lima L, et al. Over forty years of bladder cancer glycobiology: Where do glycans stand facing precision oncology? Oncotarget. 2017;**8**(53):91734

[62] Christiansen MN, Chik J, Lee L, Anugraham M, Abrahams JL, Packer NH. Cell surface protein glycosylation in cancer. Proteomics. 2014;**14**(4–5):525-546

[63] Palmigiano A, Barone R, Sturiale L, Sanfilippo C, Bua RO, Romeo DA, et al. CSF N-glycoproteomics for early diagnosis in Alzheimer's disease. Journal of Proteomics. 2016;**131**:29-37

[64] Llop E, Guerrero PE, Duran A, Barrabés S, Massaguer A, Ferri MJ, et al. Glycoprotein biomarkers for the detection of pancreatic ductal adenocarcinoma. World Journal of Gastroenterology. 2018;**24**(24):2537

[65] Brockhausen I, Yang JM, Burchell J, Whitehouse C, Taylor-Papadimitriou J. Mechanisms underlying aberrant glycosylation of MUC1 mucin in breast cancer cells. European Journal of Biochemistry. 1995;**233**(2):607-617

[66] Gils A, Pedersen KE, Skottrup P, Christensen A, Naessens D, Deinum J, et al. Biochemical importance of glycosylation of plasminogen activator inhibitor-1. Thrombosis and Haemostasis. 2003;**90**(08):206-217

[67] Lempiäinen A, Hotakainen K, Blomqvist C, Alfthan H, Stenman U-H. Hyperglycosylated human chorionic gonadotropin in serum of testicular cancer patients. Clinical Chemistry. 2012;**58**(7):1123-1129

[68] Sato Y, Nakata K, Kato Y, Shima M, Ishii N, Koji T, et al. Early recognition of hepatocellular carcinoma based on altered profiles of alpha-fetoprotein. The New England Journal of Medicine. 1993;**328**(25):1802-1806

[69] Gadroy P, Stridsberg M, Capon C, Michalski J-C, Strub J-M, van Dorsselaer A, et al. Phosphorylation and O-glycosylation sites of human chromogranin a (CGA79–439) from urine of patients with carcinoid tumors. The Journal of Biological Chemistry. 1998;**273**(51):34087-34097

[70] Jankovic MM, Milutinovic BS. Glycoforms of CA125 antigen as a possible cancer marker. Cancer Biomarkers. 2008;**4**(1):35-42

[71] Hua L, Liu Y, Zhen S, Wan D, Cao J, Gao X. Expression and biochemical characterization of recombinant human epididymis protein 4. Protein Expression and Purification. 2014;**102**: 52-62

[72] Konjevod M, Tudor L, Strac DS, Erjavec GN, Barbas C, Zarkovic N, et al. Metabolomic and glycomic findings in posttraumatic stress disorder. Progress in Neuro-Psychopharmacology and Biological Psychiatry. 2019;**88**: 181-193

[73] Wang M, Yu G, Ressom HW. Integrative analysis of proteomic, glycomic, and metabolomic data for biomarker discovery. IEEE Journal of Biomedical and Health Informatics. 2016;**20**(5):1225-1231

[74] An HJ, Kronewitter SR, de Leoz MLA, Lebrilla CB. Glycomics and disease markers. Current Opinion in Chemical Biology. 2009;**13**(5–6): 601-607

[75] Wishart DS, Tzur D, Knox C, Eisner R, Guo AC, Young N, et al. HMDB: The human metabolome database. Nucleic Acids Research. 2007; **35**(suppl_1):D521-D5D6

[76] Furukawa K, Ohmi Y, Ohkawa Y, Bhuiyan RH, Zhang P, Tajima O, et al. New era of research on cancer-associated glycosphingolipids. Cancer Science. 2019;**110**(5):1544

[77] Dehelean L, Sarbu M, Petrut A, Zamfir AD. Trends in glycolipid biomarker discovery in neurodegenerative disorders by mass spectrometry. In: Advancements of Mass Spectrometry in Biomedical Research. Springer; 2019. pp. 703-729

[78] Kitov PI, Kitova EN, Han L, Li Z, Jung J, Rodrigues E, et al. A quantitative, high-throughput method identifies protein–glycan interactions via mass spectrometry. Communications Biology. 2019;**2**(1):1-7

[79] Bonnardel F, Mariethoz J, Salentin S, Robin X, Schroeder M, Perez S, et al. UniLectin3D, a database of carbohydrate binding proteins with curated information on 3D structures and interacting ligands. Nucleic Acids Research. 2019;**47**(D1):D1236-D1D44

[80] AbuSamra DB, Argüeso P. Lectin-glycan interactions in corneal infection and inflammation. Frontiers in Immunology. 2018;**9**:2338

[81] Liu G, Neelamegham S. Integration of systems glycobiology with bioinformatics toolboxes, glycoinformatics resources, and glycoproteomics data. Wiley Interdisciplinary Reviews: Systems Biology and Medicine. 2015;**7**(4): 163-181

[82] Hucka M, Finney A, Sauro HM, Bolouri H, Doyle JC, Kitano H, et al. The systems biology markup language (SBML): A medium for representation and exchange of biochemical network models. Bioinformatics. 2003;**19**(4): 524-531

[83] Mondal N, Buffone A Jr, Stolfa G, Antonopoulos A, Lau JT, Haslam SM, et al. ST3Gal-4 is the primary sialyltransferase regulating the synthesis of E-, P-, and L-selectin ligands on human myeloid leukocytes. Blood: The Journal of the American Society of Hematology. 2015;**125**(4):687-696